America at Risk

America Burning Recommissioned

FA-223/June 2002

FEMA

AMERICA at Risk

*Findings and Recommendations
on the Role of the Fire Service
in the Prevention and Control of
Risks in America*

Table of Contents

Foreword   1
    James L. Witt, Director, FEMA

Overview of the Structure of America at Risk   3

Summary of the Commission's Process & Procedures   5

Principal Findings & Recommendations   11

Appendices

    A. Fire Control and Prevention in the U.S.: A Federal Perspective   32

    B. Status of Recommendations of the 1973 America Burning Report:
       Non-USFA or FEMA   34

    C. Members of America Burning, Recommissioned   46

    D. Commission Meeting Agendas   54

    E. Partial List of Reference Materials   59

AMERICA BURNING
Recommissioned

James Lee Witt, Director
Federal Emergency
Management Agency

FOREWORD

Foreword

One hundred years ago, American cities faced a devastating challenge from the threat of urban fires. Whole cities had become the victims of these events. Entire neighborhoods lived with the very real threat that an ignited fire would take everything, including their lives.

Today, the threat of fires is still with us. But we have done a lot to address the risk, minimize the incidence and severity of losses, and prevent fires from spreading. Our states and localities have an improving system of codes and standards; most of us are aware of the risks; our communities have everyday heroes who provide the first response to emergency calls; some of our homes and buildings have alarms or sprinkler systems; and our water distribution system for fire suppression stretches further than many imagined in 1900. We have accomplished a lot, but we have much more to do.

Our community fire departments and firefighters are at the vanguard of the long-term effort to address our fire risks. Not only are they the first responders to fire and other natural and man-made disasters, but also they have been strong advocates of effective codes and standards; they visited our schools and neighborhoods with educational material on fire risks, and they have put their lives on the line countless times. They will continue to do so. There is ample proof that the word hero is a correct attribute of our Nation's firefighters.

As the following report very clearly indicates, the success of America's fire services over the past 100 years is instructive for the strength and sustainability of America's communities for the next 100 years as well. Today, we must not only continue and reinvigorate our successes, but also expand them to include the natural and man-made threats that each of our counties, cities, towns and villages face every day – floods, earthquakes, hurricanes, hazardous material spills, highway accidents, acts of terrorism, and so much more.

As the Federal Emergency Management Agency's *Project Impact: Building Disaster Resistant Communities* has shown, community-based partnerships among local government, public safety services, businesses and residents will provide us the best set of priorities and implementation strategies, as well as the longest lasting commitments with respect to disaster prevention. That is why FEMA and national fire service organizations have formed a *Project Impact* partnership to support communities' efforts to become disaster resistant.

Project Impact depends on our first responders, our neighborhood fire departments, and without them, our communities would all be more vulnerable to disaster losses.

AMERICA BURNING
Recommissioned

Overview of the Structure of America at Risk

Overview

This report, *America at Risk,* builds on the meetings of America Burning, Recommissioned, and is based on statements, discussions and recommendations that were issued on May 3rd by the Commission as the "Principal Findings and Recommendations".

Organization of America at Risk

Chapter I provides brief information on the establishment of the Commission and the conduct of its business and meetings. Chapter II is a repeat of the *"Principal Findings and Recommendations"*. Included in this chapter is the transmittal letter of the "Principal Findings and Recommendations" from the Chairman to the FEMA Director. Relevant appendices are then included: a brief history of the Federal sector's involvement in support to the Fire services since the late 1940's; a compilation of the responses received from Federal and private non-profit agencies on the status of follow up on recommendations of the 1973 report, *America Burning* that were directed toward them; a list of the members of *America Burning, Recommissioned* with information on their relevant background; and a compilation of the commission's meeting agendas.

Acknowledgements

The Commission expresses its sincere gratitude to all of the organizations and individuals that provided comments, offered assistance, submitted materials, spoke to them during their meetings, attended the meetings, and generally made the Commission's task easier.

At several places within this report, mostly in Chapter II, are quotations from comments that FEMA received at its *America Burning, Recommissioned* web site on issues confronting America's fire services, and recommended approaches for addressing those issues. We have attempted to use a representative sample in order to illustrate the sentiments and beliefs of the fire services community with respect to the findings and recommendations. As stated in Chapter I, Commission members availed themselves of these comments in the development of their report.

The pictures and images used within the report were provided by the USFA photo library; the Prince George's County, Maryland Fire Department; the International Association of Fire Chiefs; FEMA photographer Andrea Booher and FEMA's Office of Media Affairs. They are intended to illustrate the story of *America at Risk* and the front line role that firefighters carry out in that daily challenge.

AMERICA BURNING
Recommissioned

Summary of the Commission's Process and Procedures

COURTESY PRINCE GEORGE'S COUNTY F. RE DEPT.

Summary of Commission's Process and Procedures

Origin of America at Risk

In the late summer of 1999, the FEMA Director formally recommissioned *America Burning.*

America Burning, Recommissioned was a response to the recommendation of the Blue Ribbon Panel, which had provided its report to the Director early in October 1998. This expert panel had been assembled by the FEMA Director to give him an assessment of concerns and issues with respect to the ongoing work of the USFA, and to obtain recommendations on how to improve the effectiveness of this critical component of the FEMA organization. The underlying rationale of the members of the Blue Ribbon Panel, in their recommendation that *America Burning* be recommissioned clearly was that *America Burning,* in 1973, had provided important foundations and focus to the management of fire risks in America. It had resulted in the establishment of USFA's predecessor organization, and it had created a blueprint for its activities. If such efforts could be adapted to current conditions, beneficial progress and results for the fire services as a whole would follow.

The FEMA Director acknowledged the importance of the Blue Ribbon Panel's recommendations, and the Panel's reasoning for it. He also articulated his exact purpose for recommissioning *America Burning* in his letter, and in his visits with the Commission members during their meetings in Washington, D.C. In his letter to the members, The Director wrote that:

> *"…(the commission will fulfill an) essential role in the initiation of a much needed, and long-awaited, national effort to continue tangible reductions in our country's losses to fires. Equally importantly, it will also provide a critical framework for the evolving role of the fire services in the safety and sustainability of today's American communities… Your panel will recommend an approach to an updated and renewed vision for the fire service community."*

In its first meeting, the members of *America Burning, Recommissioned* determined that, in order to carry out their responsibilities, their considerations and eventual recommendations to the Director would have to go beyond the boundaries of the fire risk alone. *America Burning* in 1973 had anticipated some of the forthcoming challenges to the fire services. They had, for example, correctly identified the importance of emergency medical services. Yet to the members of *America Burning, Recommissioned*, there had been developments even beyond what had been forecast in 1973. The establishment of FEMA, the growth of the emergency management community as a profession, the increase in disaster losses in America, and other factors, had dictated a different context for the fire services. Still, the members suggest that it is not a context that is unfamiliar, or one that requires difficult adjustments for the fire services. Indeed, it is a framework within which any firefighter and any fire department can be effective. America's communities have a range of disasters with which they are confronted. America's firefighters are still communities' first line of response to this range of disasters. Accordingly, *America at Risk* was determined to be the correct title and orientation of the Commission's report.

Commission Procedures for Obtaining Outside Input and Comment

In the last section of this report, each of the Commission's agendas are reproduced so that those organizations that made oral or oral/written presentations to the Commission can be readily noted. During the public portion of the agendas of their meetings, the Commission members unanimously expressed the desire that they take every opportunity to hear varying points of view on the multitude of issues that today confront the fire services. Many groups gave freely of their time and resources to attend Commission meetings, develop presentations, and be responsive to the Commission's subsequent discussions and inquiries. Although the Commission had a short time frame in which to conduct four meetings, formulate options for its written findings and final recommendations, a broad cross section of interests was invited to make presentations.

In addition, the Commission considered obtaining input from the fire services community and others through various other means:

★ Mail (both traditional and electronic or "email") comments were numerous and extremely useful to the members. Over 50 pieces of written "mail" correspondence were received that either provided comments on relevant subjects, provided enclosed publications or other written material for the Commission's information, or offered assistance to the Commission in its work;

★ In order to take advantage of the growing use of personal computer technologies, the Commission established a web site so that comments could be received from virtually everywhere. Nearly 200 comments were received from over 45 States and two foreign countries. Excerpts from the submitted comments are included in various locations in this report; and

★ Had there been more time, the Commission would have conducted "field" meetings in various U.S. regional locations. This approach may yet be possible to develop detailed and grass roots oriented strategies for carrying out some of the Commission's recommendations.

Commission Procedures for Formulation of Meeting Agendas

The members of *America Burning, Recommissioned* held four meetings of two days each in Washington, D.C. The Chairman, who used recommendations from the FEMA Director, USFA leadership, and the Commission support staff, formulated the agenda for the first meeting. Thereafter, Commission members provided the principal input to identify those topics that they believed were important to cover in their limited meeting time frames, and to identify those organizations or individuals that should be asked to provide topical information and data. Invited presenters were given broad guidelines within which to work.

Publications, Background Materials or other Reading Materials

Commission members reviewed a large set of materials (see Appendix E) provided by the FEMA support staff, the members themselves, and interested and concerned organizations or individuals who provided materials through the regular mail.

Preparation of the Commission's Findings and Recommendations

As the Commission formulated its principal message for the FEMA Director, it also came to the conclusion that a crucial audience would also be the U.S. Congress. In recognition of this importance, the members decided to utilize the occasion of the Congressional Fire Services Caucus meetings of May 2000 as the date by which it would issue its Principal Findings and Recommendations. On May 3, that report was issued. The Commission also directed that its full report, America at Risk, be prepared on the foundation of the May 3rd document, and issued as soon as possible.

Preparation of the Full Report: America at Risk

In June 2000, a five-member sub-group of the Commission met at USFA in Emmitsburg, Maryland to begin preparation of the full report. Using the *Principle Findings and Recommendations* as its foundation, the sub-group's primary objective was the preparation of additional text that would substantiate each of the findings in the May 3rd document and describe the consensus, residing either inside or outside the fire services community that might underlie the recommendations. In addition, there are issues confronting the fire services that were addressed generically among the principle findings and recommendations, but that warranted special attention.

Staff, which supported the efforts of the members of the Commission, spent considerable time after the June sub-group meeting researching further the existing written and electronic databases that underlay the findings of the May 3rd report. As a result, there are, for example, data to support aggressive implementation campaigns for the Commission's recommendations with respect to prevention. In those instances that data can be employed now to accelerate achievement of the Commission's recommendations, FEMA/USFA is strongly urged to do so. Moreover, the effort to collect quantitative data should be continued. However, the Commission recognizes that encyclopedic data for all the recommendations may not exist, or, in several instances, will be some time in compilation.

Specific issues that the sub-group identified as warranting specific follow-on attention include:

★ The hazard and risk management issues and requirements associated with fires located generally at the point of **urban/wildland interface.** The requirements can be separated into a) effective (for realty loss prevention) and safe (for firefighters) response and suppression capability and b) the reduction of existing wild fire hazard and the avoidance of new wild fire risks through effective mitigation techniques. The issues associated with these requirements include appropriate forest management, sound land use and community growth principles, training and equipment, research and data, and partnerships to overcome obstacles such as resources, awareness, and long-term commitments;

★ The societal and criminal issues associated with the risk of **arson**, and particularly the challenge of dealing with **juvenile fire setters.** In large part, due to the convergence of the arson fire issue with the arson crime issue, there are substantial data on the arson problem in America that can be, and indeed have been, used to develop and implement effective arson prevention strategies. Notable effective strategies for addressing the arson and the juvenile fire setter problems have been carried out in the City of Utica, New York and Orange County, California, respectively; and

★ The management and customer service issues associated with the evolution of the fire services not only as a profession, but also as a business. This issue is an umbrella for several of the specific matters addressed in the May 3rd report – such as diversity, health of firefighters, and the appropriate implementation of emergency medical services – as well as with issues indirectly referred to in that report – such as the ability of community fire services to attract financial and human resources; the capacity of the fire service to effectively work within America's evolving societal circumstances, and the expansion of the fire services into hazard and disaster management areas that are increasingly critical to the sustainability of today's communities.

AMERICA BURNING
Recommissioned

Principal Findings & Recommendations

COURTESY CHIEFS

THE BERNSTEIN LAW FIRM, PLLC
1730 K STREET, N.W.
WASHINGTON, D.C. 20006-3868
TELEPHONE (202) 452-8010
FACSIMILE (202) 296-2085

GEORGE K. BERNSTEIN
CARYL S. BERNSTEIN
ROBERT E. SHAPIRO

NEW YORK OFFICE
30 ROCKEFELLER PLAZA
NEW YORK, NY 10112
(212) 957-7604

May 2, 2000

Honorable James Lee Witt
Director, Federal Emergency Management Agency
500 C Street, SW
Washington, DC 20472

Re: America Burning Recommissioned

Dear Director Witt:

In creating this Commission, you directed us to recommend an updated approach and framework for the evolving role of the fire service with respect to fire and other hazards. You also asked us to review the 1972 *America Burning* Report in light of the 1998 review by the "Blue Ribbon Panel" of the U.S. Fire Administration ("USFA"), which "identified three core deficiencies that are undermining the effectiveness of" the USFA. You have emphasized that your major concern is that the Commission identify the means by which fires can best be prevented and loss of life and property reduced.

America Burning was considered the seminal effort in systematizing our nation's efforts to address the fire hazard and the resultant loss of life and property. It was greeted with praise by all elements of government and the fire community. However, more than one-third of the *America Burning's* recommendations have not been implemented and more than half were only partially implemented. During the intervening 28 years there was no systematic effort to track the implementation of these recommendations.

A prime example of the failure to adequately implement *America Burning* has been the unwillingness of the federal government to fund the USFA and ancillary programs at anywhere near the recommended level of $153 million. The initial funding in 1980 was only $24 million ($45,130,000 in year 2000 dollars) and twenty five years later, for fiscal year 2000, it is only $42,982,000, an actual decrease of 4.7 percent, reflecting inflation.

The lack of substantial funding to implement America Burning speaks volumes about the low priority that all segments of government — federal, state and local — assign the fire hazard compared to other areas of public safety. The failure to adequately fund fire prevention and response, in general, and the USFA, in particular, has resulted in continued loss of lifeand property at levels that would otherwise have been substantially reduced.

Although deaths from fire have fallen from 7,395 in 1977 to 4,035 in 1998, those lost lives are not acceptable. One-hundred and tenfirefighters died in the line of duty in 1998. Despite the occasional periodic sympathy generated by tragedies such as the loss of six firefighters in Worcester, Massachusetts in December 1999, fire deaths receive little nationwide attention and sparse legislative and funding response. Since the Worcester fire, USFA has been notified of 40 firefighter fatalities and estimates that an additional 1,800 civilian lives have been lost from fires.

It is also unfortunate that, despite the current involvement of the fire services, often as first responder on the scene, in such areas as emergency medical services, hazardous materials and acts of domestic terrorism, the fire services have not been granted the additional wherewithal to carry out these new responsibilities.

The Commission concurs with the conclusion of the Blue Ribbon Panel that the USFA's operations suffer from deficiencies in "leadership" and "communication." We believe that such deficiencies, which FEMA has begun to address, are derivative of the national lack of priority given to support of the fire services. As to the Blue Ribbon Panel's finding of deficiencies in "resource management," we agree with the Panel that "At no time since 1974 has the USFA had the resources it needs to address this nation's fire problems with sustained impact." We believe that many of the other problems of "resource management" are also derivative of inadequate funding over the years and the attempt to accomplish too many tasks with too little money.

Ways to reduce fire losses and deaths are neither unknown nor arcane. The primary way and the goal of any effort in this area must be to prevent fires in the first place. Smoke detectors and alarm systems in homes and commercial buildings have already proven their worth in alerting occupants and saving lives. Sprinklers are acknowledged as the most effective tool in immediately suppressing fires, minimizing damage and saving lives.

Unfortunately, few jurisdictions require sprinklers in private dwellings or existing commercial and residential structures, or even in existing schools, nursing homes, hospitals or other places of public assembly, except in the event of major reconstruction. Moreover, the federal government sets a poor example by its failure to require sprinklers in assisted public housing or in existing federal buildings.

Although our Commission appreciates the concern of this Administration with the current fire situation, as represented by the recommissioning of *America Burning*," we are concerned, based on prior governmental inaction, whether there is the will and commitment at the federal, state and local levels to fund and implement actions that are essential to reduce losses from fire and other hazards.

Until the USFA is empowered by funding and staffing to become the leader in our nation's firefighting efforts, unless the fire services are adequately funded, and unless local communities enforce known fire preventive and suppression measures, the establishment of this Commission and its efforts to develop recommendations in the areas of leadership, research, training, communication and public education, sound coode development, partnership between the public and private sectors, and above all, in mitigation, will be an exercise in futility.

If the number of fires and resultant losses are to be reduced, there must be a concerted and consistent effort among not just the fire services but, as recently observed by three of those involved in the original America Burning Report and its initial implementation, other stakeholder groups as well, including "city and county managers, mayors, architects, engineers, researchers, academics, materials producers and the insurance industry" as well as the general public.

Although members of the Commission have diverse backgrounds and concerns, they have subordinated parochial interests to the development of recommendations to prevent fires and reduce loss of life and property from fire and other hazards. I thank you for the privilege of working with this distinguished group.

The attached Findings and Recommendations summarize the conclusions of the Commission. We will shortly forward to you the complete Report. A number of our recommendations are not new and reflect the foresight and wisdom of the authors of the original America Burning as well as other studies since then. The Commission believes that had those past recommendations been more fully implemented, there would have been less need for this report. We hope that sufficient action will follow this Report so that twenty-five years from now another *America Burning, Recommissioned* will be unnecessary.

Sincerely,

George K. Bernstein,
Chair

Introduction

To a great extent, the fire problem in America remains as severe as it was 30 years ago. If progress is measured in terms of loss of life, then the progress in addressing the problem, which began with the first *America Burning* report in 1973, has come to a virtual standstill. The "indifference with which Americans confront the subject" which the 1973 Commission found so striking continues today. Yet today's fire departments, rescue squads, emergency service organizations and other first responders face expanded responsibilities and broader assignments than traditional structural fire response and suppression. To address this dilemma, the Director of the Federal Emergency Management Agency recommissioned *America Burning* in late 1999.

Since its formation, the Commission conducted four meetings and in addition to its deliberations, heard testimony and received input from approximately 30 individuals and groups, received written submissions from over 50 parties and established a website on which 191 responses were filed.

The Commission reached two major conclusions:

(1) The frequency and severity of fires in America do not result from a lack of knowledge of the causes, means of prevention or methods of suppression. We have a fire "problem" because our nation has failed to adequately apply and fund known loss reduction strategies. Had past recommendations of America Burning and subsequent reports been implemented, there would have been no need for this Commission. Unless those recommendations and the ones that follow are funded and implemented, the Commission's efforts will have been an exercise in futility.

The primary responsibility for fire prevention and suppression and action with respect to other hazards dealt with by the fire services properly rests with the states and local governments. Nevertheless, a substantial role exists for the federal government in funding and technical support.

(2) The responsibilities of today's fire departments extend well beyond the traditional fire hazard. The fire service is the primary responder to almost all local hazards, protecting a community's commercial as well as human assets and firehouses are the closest connection government has to disaster-threatened neighborhoods. Firefighters, who too frequently expose themselves to unnecessary risk, and the communities they serve, would all benefit if there was the same dedication to the avoidance of loss from fires and other hazards that exists in the conduct of fire suppression and rescue operations.

Fire protection expertise can be shifted from reaction and response and used for prevention activities.

A reasonably disaster-resistant America will not be achieved until there is greater acknowledgment of the importance of the fire service and a willingness at all levels of government to adequately fund the needs and responsibilities of the fire service. The lack of public understanding about the fire hazard is reflected in the continued rate of loss of life and property. The efforts of local fire departments to educate children and others must intensify. Without the integrated efforts of all segments of the community, including city and county managers, mayors, architects, engineers, researchers, academics, materials producers and the insurance industry, as well as the fire service, there is little reason to expect that a proper appreciation of the critical role played by the fire service will materialize, in which case the necessary funding will continue to be lacking.

These conclusions underlie the findings and recommendations that follow.

FINDING #1

Implementation of Loss Prevention Strategies

The strategies and techniques to address fire risks in structures are known. When implemented, these means have proven effective in the reduction of losses. The tragic reality, however, is that existing and effective strategies have not been funded adequately by the Congress or state and local governments, nor have they been aggressively advocated by the United States Fire Administration (USFA) and other fire service constituencies. As a consequence, America today has the highest fire losses in terms of both frequency and total losses of any modern technological society. Losses from fire at the high rate experienced in America are avoidable and should be as unacceptable as deaths and losses caused by drunk driving or deaths of children accidentally killed playing with guns.

Comprehensive proposals to address structural fire risks were contained in the recommendations of the 1973 *America Burning* report. The wisdom of these recommendations was acknowledged by the Congress and the Administration in the enactment of the Federal Fire Prevention and Control Act of 1974 (the "1974 Act"). However, FEMA and the USFA have not pursued many of the preventive measures authorized by that statute; the Congress has not appropriated the funds necessary to carry them out; they have not been adequately advocated by USFA; and if implementation is the test, they have not been widely accepted by the fire service-at-large. Since 1974, successful approaches for implementing mitigation measures have been developed, but have not been incorporated in comprehensive programs to reduce structural fire loss. In addition, FEMA has not applied to the fire problem those lessons which it learned with respect to other natural hazards, including earthquake, flood, and hurricane and has failed to exercise all of its powers under the 1974 Act.

COURTESY PRINCE GEORGE'S COUNTY FIRE DEPT.

Recommendations:

The Congress should increase its involvement in fire loss prevention in America and exercise more fully its oversight responsibilities under the 1974 Act. The Congress should also appropriate for the fire problem appropriate resources commensurate with those it provides to community policing or highway safety. FEMA should exercise its full authority under the 1974 Act and should apply to the fire hazard the same prevention emphases and strategies that it has applied to other natural hazards, the Agency's objective being an all-risk, multi-hazard loss prevention program.

FINDING #2

The Application and Use of Sprinkler Technology

The most effective fire loss prevention and reduction measure with respect to both life and property is the installation and maintenance of fire sprinklers. If the focus is limited to prevention and reduction of the loss of life, smoke alarms are also extremely effective. However, the use of sprinklers and smoke detectors has not been sufficiently comprehensive.

Recommendations:

FEMA/USFA should develop a long-term implementation strategy for fire sprinklers and smoke alarms. The plan should include the following implementation aspects:

★ The approach should be community based;

★ No tactic or strategy should detract from the requirement for sprinklers. Smoke alarms (or other measures) should always be the locality's second option as a loss reduction measure;

★ Exploration of the technical, economic and practical aspects of utilizing alarm and sprinkler systems that provide automatic notification to a firehouse. These systems should be professionally maintained and monitored;

Sprinklers should be installed in all new construction and phased in incrementally for varying types of existing construction.

COURTESY USFA

COURTESY PRINCE GEORGE'S COUNTY FIRE DEPT.

★ Confirmation of the accuracy of the belief that the appropriateness of the emplacement of sprinklers and alarms may be based on rural and urban distinctions, and whether other distinctions such as residential construction, commercial construction and critical facilities may also be appropriate;

★ The plan should distinguish between requirements for new construction and existing construction.

The plan should articulate actions that will result in:

★ Improved use of financial incentives;

★ Government leadership in including fire safety measures in its own buildings, and in those that it helps construct or for which it provides any form of financial assistance or guarantee;

★ Prioritization standards in the retrofit of existing buildings based on risk to the public;

★ A national public awareness and education campaign;

★ Participation of the private and academic sectors;

★ Improvement of technologies and lowering of costs;

★ Inclusion and enhancement of fire safety requirements in model building codes and standards; and

★ The plan should complement communities' actions to address all their hazards. For example, the ability of a community to address fire hazards should not be compromised by an earthquake event that ruptures sprinkler systems.

FINDING #3

Loss Prevention Education for the Public

The most effective way to reduce the loss of life from natural and man-made disasters is through a multi-hazard mitigation process that addresses all the hazards a community faces. Currently, FEMA has begun a community-based, all-risk program entitled, Project Impact: Building Disaster Resistant Communities. The National Fire Protection Association (NFPA) has also begun a program entitled *Risk Watch*, which includes many of the approaches of *Project Impact.*

Too many fires are caused by carelessness and ignorance of principles thought to be obvious. Education about the fire hazard should reach children who are responsible for so many accidental fires. It has been the experience of the fire services that schools are one of the best venues for firefighters in providing safety information to children and young adults. Thus, the fire services can play an important role in developing mitigation and prevention awareness programs through and in neighborhood schools. Our youngest citizens would then have the opportunity to appreciate, convey to parents and even implement life saving initiatives. A unified fire prevention curriculum should be written, tested and validated by education specialists to provide a complete package for citizens.

Recommendations:

These mitigation programs should be combined in a unified all-hazard learning curriculum and implemented nationally by community and neighborhood fire services in all levels of the local school systems. Fire departments should be encouraged to spend even more time in reaching out to children in schools and other venues. By providing a community-based and complete package to educators, fire service representatives can work from the same baseline of information to ensure that a consistent message is sent nationwide.

In addition, effective public service commercials demonstrating the risks and avoidance techniques for fire and other hazards should be pursued. The success of such federal initiatives as seat belt use hold great promise for public education on the issues of fire.

Further findings and Recommendations with respect to the issues of public education and awareness will be presented in #7 below.

FINDING #4
The Acquisition and Analysis of Data

Collection and analysis of meaningful data is critical in order to address the fire problem with respect to civilian and firefighter casualties. Analysis of data provides a basis for direction and prioritization to initiatives discussed herein.

A large quantity of data exists. However, the strategic quality and significance of much of these data are not apparent or have been questioned. The Commission is unaware that the data collected are effective for advancing or achieving the prevention goals of the fire prevention and services community. In addition, there is no central center or focus for the analysis of data that are collected. It is not clear whether the current National Fire Incident Reporting System has cost inefficiencies with respect to data overlap or is providing corroborating data, whether there is under-utilization for data analysis purposes, or whether there is national applicability of data that are present.

The fire and emergency services community needs a central, national data center on which to rely for the collection and analysis of data. The analysis of data should underlie funding and public policy decisions that address problems or issues identified in the data. For such a center to be effective in this role, all regions and states should participate in and contribute to the collection of relevant data. Data that are collected by any institution or organization should have utility, in both form and substance, with the data that are collected by other entities. The data received by the center should be available to outside sources.

COURTESY NIST

This compatibility of data is critical and reflects the fact that there are and will continue to be many entities collecting relevant and useful data. In the future, the mutual reliance of these differing participants should become emphasized, to the point that their work has shared objectives, goals and activities.

Complete and encyclopedic data are not a pragmatic requirement in the achievement of all fire goals and objectives. The fire and emergency services community should be able to rely on state-of-the-art statistical sampling techniques to define community problems, jurisdictional challenges and the issues confronting the nation. This will provide a more efficient method of defining risk reduction efforts and formulating public policy.

As a practical and political matter, adequate financial resources will not always be provided by the Congress. However, there are strategies that can be implemented to both supplement federal resources and leverage additional resources into the data collection and analysis category.

Recommendations:

FEMA/USFA should develop a plan to effect appropriate data collection and analysis. The plan should include a reconciliation of existing FEMA data systems, as well as identifying adequate levels of funding needed to revive both data collection and data analysis and use. Resources to achieve the plan should also be identified and pursued. The plan should include the following actions and aspects:

★ FEMA and USFA should facilitate or initiate working partnerships that further efforts to institutionalize the compatibility of data on the part of allied organizations and agencies. The all-hazards aspect can also be reflected by including organizations such as the Insurance Services Office (ISO), the National Fire Protection Association (NFPA), the Bureau of Alcohol, Tobacco and Firearms (ATF), the U.S. Geological Survey, the National Oceanic and Atmospheric Administration, and others.

★ FEMA/USFA should also have state government partners in the collection of data. To this end, FEMA/USFA should encourage state collection of data by providing financial incentives through the grant process.

★ There should be a one-time examination of the practicality of developing a statistical sampling model that can be utilized by the various regions, states and local communities as appropriate.

★ For the national data center to be effective and efficient and to be adequately funded, there should be a transparent process for the setting of the agenda for the center so that problem-focused analyses can be prioritized and shared with its partners. In some instances, it may also be feasible for such partners to perform needed analyses on their own initiative.

★ After-action data, which is not currently collated, should be collected and analyzed by the center. Such data should identify the pre-event activities, (e.g., preventive actions, codes or standards, training) and response activities (including equipment, techniques, etc.) proved most effective.

Improvements Through Research

Research on the science of fire, fire behavior, the suppression and extinguishing of fire, and fire service operations is inadequate. Valuable investigation is currently being conducted in Federal Agencies, such as the Consumer Product Safety Commission and the National Institute of Standards and Technology (NIST). However, this research is not coordinated, prioritized or focused on identified problems. Valuable research is also ongoing at many of the Nation's colleges and universities and there is also a private sector component of research into fire and emergency services issues that may contribute to a national agenda.

The transfer of research results into practice can also be improved. First, technology transfer is not facilitated by the Federal sector in any efficient manner. The private sector has extensive relationships with most of the fire research community, and it is these (informal) relationships that seem to result in most of the technology transfer. Conversely, it is equally important that the end-user, the practicing fire and emergency services community, be able to communicate to the research community its problems and issues, and to directly influence research priorities.

The Commission considered this latter aspect when it evaluated the lack of empirical research results to support changes to model codes and standards. Many such changes have been based on the "equivalency" concept, and assume that the building owner is making offsetting structural improvements that obviate or reduce the need for previous fire retardant code requirements. While certain fire loss prevention components of the construction may have been researched, there reportedly has not been research into the impact of these decisions on the safety of firefighters who, if a fire did occur, would have to enter the building to conduct manual fire suppression activities.

Because of the all-hazard responsibilities of the fire services and emergency management community, the number of researchers involved grows significantly, the prioritization of needs compounds, and coordination and technology transfer becomes even more important.

The roots of the current lack of coordinated research effort may lie in the separation of certain functions between FEMA/USFA and NIST (then the National Bureau of Standards) when FEMA was formed in 1979. However, as

COURTESY NIST

COURTESY PRINCE GEORGE'S COUNTY FIRE DEPT.

indicated elsewhere in this report, the character of the fire and emergency services has changed dramatically since the Fire Prevention and Control Act of 1974. Therefore reverting to the earlier research arrangements contemplated in the Act would not be appropriate.

Recommendations:

FEMA/USFA should take a leadership role in setting agendas for research into fire and other risks for which the fire and emergency services community have responsibility. As a first step, a reasonable set of priorities should be established for fire issues. Research agendas should be set with significant user input and influence. In addition, partnerships among NIST and other governmental, university, international and private research organizations can be utilized to develop research agendas that include issues connected with building codes and standards.

The agendas should be followed by the development of an implementation plan that specifies the organization, institution, or private sector partner responsible for the completion of the research. Resource needs should also be identified and adequate funding should be pursued vigorously.

FEMA/USFA should not allow the development of an agenda for "fire" to become a single-hazard issue, for two important reasons. First, as stated elsewhere in the Commission's findings and recommendations, because of the all-hazard nature of their responsibilities, the fire services have clearly become the fire and emergency services. Secondly, FEMA conducts or participates in other hazards programs – e.g., hazardous materials, terrorism, and earthquake and other natural hazards – that include research within the programs' activities. Within a reasonable time, the "competing" agendas of these programs should be coordinated and ultimately integrated.

With respect to the critical subject of technology transfer, the Commission understands that FEMA/USFA already does important work to make research results available, but believes that other initiatives can be pursued in order to make the process more efficient and expedient. Trade press columns, conferences or conventions, and partnerships with public and private sector organizations can be utilized to accomplish the goal. In addition, the new technologies and other results of relevant research should be incorporated into the courses and documents offered at the National Fire Academy.

FINDING #6

Codes and Standards for Fire Loss Reduction in the Built Environment

There should be an active and aggressive approach by FEMA/USFA in the utilization of building codes and standards for construction in order to prevent or reduce fire losses. To date, there has been success in the use of codes and standards. However, the success must be accelerated and intensified.

The adoption and enforcement of those codes and standards for construction or rehabilitation that affect fire safety (as well as safety for all hazards) must be extended. The Commission's discussions focused on the need to address more of the residential losses, the potential losses in existing (or new) critical facilities, and the losses in structures that contain vulnerable populations (e.g., retirement homes). Changes to model codes and standards that reflect research that validates the revisions would thus provide the technical basis for local and state adoption and enforcement of measures that address local and state risk management priorities.

The need for emphasis on residential construction is born out by statistics. For the most recently compiled year, 1997, there were 552,000 structure fires in the United States. Almost three-quarters of structure fires occurred in residential properties including homes, hotels, motels, rooming houses and dormitories. Fifty-five percent (55%) or 302,500 were in one- and two-family homes and seventeen percent (17%) or 93,000 occurred in apartments. The largest number of civilian deaths occurred in residential buildings. Eighty-three percent (83%) of the 4035 total civilian deaths occurred in home structure fires – with sixty-seven percent (67%) or 2700 in one-and two-family homes.

There are major improvements in the effectiveness and efficiency of the U.S. codes and standards system that would be realized from the joint efforts of appropriate organizations from the fire, emergency services, and building communities. The Fire Prevention and Control Act of 1974 gives USFA an important role and authority to effect this integration (from the fire services point of view) but that authority has not been exercised. The safety of new buildings, and the ongoing inspection and enforcement of those safety provisions in existing buildings, would be improved by this integration.

Recommendations:

The USFA should review its authority under the Fire Prevention and Control Act of 1974 in order to identify those activities it could support, but currently does not, with respect to building codes and standards. These activities would include:

★ The development and promulgation of a set of performance standards for buildings, with respect to fire hazards and risks, against which model codes and standards can be measured for equivalency. The participation and consensus processes used by FEMA to develop such standards for seismic vulnerability in buildings may serve as a useful paradigm;

★ The active involvement of the fire services community in the consensus process of model code promulgation gives the drafters the benefit of real experience in the prevention and suppression of fire and to ensure that the current trend towards "equivalency" does not unintentionally put firefighters at additional risk;

★ The development of training courses on the enforcement of building and fire codes in new and existing buildings at the National Fire Academy (NFA) that can be handed off to state and local governments. In addition, USFA should utilize its present and emerging academic partnerships with colleges and universities that have architectural and engineering programs to ensure that fire safety inspections

One vital change I see is a closer relationship between fire service and building safety organizations.

and code enforcement are a part of the curriculum; and

★ The identification of improved or enhanced insurance incentives for community-based fire loss prevention measures and homeowner loss reduction implementation, especially fire sprinklers and alarms.

FINDING #7

Public Education and Awareness

There is wide acknowledgment and acceptance that public education programs on fire prevention are effective. The reduction of the number of fire deaths since the first America Burning report is due to a number of factors, including increased awareness that fire is not an inevitable tragedy. As with efforts to prevent or reduce losses from other hazards, such as earthquake, flood and hurricane, public education will not be totally effective on its own. Nevertheless, no prevention effort can succeed without a public education component. Social marketing techniques appear to have the greatest likelihood for success on fire issues since they seek to change the way people think and make decisions.

A public education approach should be mindful of two essential elements: first, the public education must make the target audience aware of the hazards on both an intellectual and emotional level. Second, the target audience must receive and accept the message that the hazard or problem is within its control.

Recommendations:

FEMA/USFA should develop and support a public awareness campaign strategy that includes the following features:

★ Measurable results, goals and objectives;

★ Targeting high-risk areas with concentrated efforts and appropriate messages on public education and fire prevention;

★ Use of existing community resources (e.g., schools, community groups and activities, houses of worship, and social, medical, and other education services), to deliver the message to audiences already in place;

★ The development and utilization of private sector partnerships with enterprises that have investments in the reduction of fire losses, such as insurance companies, both property casualty and life and health;

★ Though instituted at the national level, capable of being carried out at the local level;

★ Training to prepare fire officers to deal with the media - for public information, education, and relations; and

★ A multi-hazard approach that advances prevention and safety messages for all of the risks which fire departments respond to and address and that educates about the multifaceted approaches involving code enforcement, construction standards, education, and enhanced technology usage such as sprinklers and smoke detectors.

FINDING #8

National Accrediting and Certification

Fire training and education in the United States remains disparate and unequal. There are recognized standards, accreditation and certification processes, but the country still lacks a nationally recognized system envisioned by the 1973 Commission. Firefighters and officers trained in one state may have to repeat all of their training before they can serve as a firefighter in another state. Colleges and universities do not have a model curriculum upon which to base their degree programs.

Recommendations:

FEMA/USFA/NFA should begin the process of establishing a system of training and education that is nationally recognized and reciprocal among the states. Participation in the system by state, local and college-based training systems should be voluntary, but USFA/NFA should provide incentives for participation.

In order to enhance distribution of USFA/NFA training, state fire training systems should be authorized to deliver USFA/NFA campus-based programs, use USFA/NFA instructors, and issue USFA/NFA certificates to students. Courses should be delivered at times and places convenient to the state systems. Though independent, state training systems should be considered extensions of the USFA/NFA delivery system.

COURTESY PRINCE GEORGE'S COUNTY FIRE DEPT.

USFA/NFA should establish a peer-review process by which courses developed by state training systems are reviewed for endorsement by the USFA/NFA. These endorsed courses should be shared among state and local training systems. The endorsement process will increase the number of courses available to state training systems, provide local systems with courses on subjects that meet local needs, and begin the process of establishing a national system of training and education envisioned by the original America Burning Commission.

The process by which courses are "handed off" to state training systems should be re-engineered. The focus should be on getting USFA/NFA developed courses into state and local training systems more quickly and involving instructors in the course revision/edits process.

The number of technology-based courses should be increased. CD and Internet technologies should be utilized to reduce the amount of paper based materials currently printed, stored and shipped to state and local training systems.

As an additional incentive to the development of courses and the establishment of a reciprocity system, performance-based training grants should be made to state training systems that permit them to deliver not only USFA/NFA developed courses but also courses that have met USFA/NFA endorsement criteria for off-campus delivery.

Participation by colleges and universities in the national fire prevention efforts should be expanded and a group of colleges and universities should be convened to help design a model curriculum.

FINDING #9

Firefighter Health and Safety

It is evident that a key element in the reduction or prevention of the loss of life and property at a fire emergency is a properly organized, staffed and deployed fire department. A fire emergency is a time sensitive and labor intensive task. Many fire departments in the United States today do not have the capacity to provide all the requisite functions required for an initial first alarm response in a timely manner.

Moreover, as noted elsewhere in the Commission's Findings and Recommendations, firefighters respond to all hazardous incidents in a community, not only fires. Firefighters respond to over a quarter of a million hazardous material incidents each year in the U.S. They are tasked with protecting the public during and after an incident involving weapons of mass destruction. They perform rescue operations in a multitude of circumstances ranging from natural disasters to voluntary endangerment by ultimate sports participants. Training for these operations is frequently substandard where it exists. Worse, it is usually absent in key areas such as safety for firefighters from hazards external to the incident site (e.g., high-speed traffic at the site of a highway accident) and responder health and safety with respect to the causative hazard (e.g., appropriate equipment for response to a hazardous materials incident).

COURTESY PRINCE GEORGE'S COUNTY FIRE DEPT

Fire departments are also now called upon to provide emergency medical response at various levels from first responder to advanced life support and transport. Existing EMS response systems, including some under the fire service, often provide inconsistent emergency medical response coverage, are understaffed and undertrained, and do not deploy and arrive at medical emergencies within medically accepted response times.

Thousands of fire fighters and emergency medical personnel lack rudimentary medical evaluation and wellness/fitness programs that can dramatically work to ameliorate the negative effects of emergency response and toxic exposure. Too many fire fighters and paramedics suffer from cancer as the result of chronic exposure to toxic products of combustion and the numbers continue to increase. Additionally, each year more firefighters are exposed to infectious diseases during the provision of basic and advanced life support in uncontrolled, emergency environments.

Protective clothing and equipment utilized by fire fighters and emergency medical personnel are not always properly selected, used, and maintained. Inferior products are still sold to and procured by fire departments.

Recommendations:

Communities that fund fire departments to respond to fire emergencies within their jurisdiction should be fully cognizant of the capacity of the department in terms of its deployment capability, including structural fire response, special operations and hazardous materials response, and emergency medical response. Fire departments should be evaluated based on their effectiveness, efficiency and worker safety. The decision of the jurisdictions' level of service should be based on technically, scientifically and medically sound criteria for organization, staffing and deployment of such services. Fire fighters and emergency medical personnel should be selected for the job based on consistent medical and performance standards.

All fire departments should provide protective clothing and equipment as well as specific training for the prevention of occupationally acquired infectious diseases, cancer, heart disease and other occupationally related diseases. Such clothing and equipment must provide continual protection during its use against the hazardous conditions encountered during fire fighting, emergency medical and special operation functions.

FEMA/USFA should directly support or advocate the development of nationally applicable assessment and evaluation systems on the full range of operating capabilities and capacities of public fire departments. Such systems should be adopted, and if necessary promulgated by the appropriate

federal agency. The evaluation system should be based on the minimum functions and tasks required for fire, medical or other emergencies, as well as the minimum response times required to deliver such services, and should measure the effectiveness and efficiency of public fire suppression, emergency medical services, and special operations delivery in protecting both the public and the occupational safety and health of fire department employees.

FEMA/USFA and other appropriate federal agencies should encourage all fire departments to adopt a standard operating procedure addressing safe incident-site staffing that includes accountability and teams for fire fighter rescue.

Fire departments should provide a wellness/fitness program to maintain the medical, physical and behavioral health of all personnel. The federal government should provide funding for fire department adoption of fire fighter wellness/fitness programs based on the Wellness-Fitness Initiative and the Candidate Physical Ability Test of the International Association of Fire Fighters and the International Association of Fire Chiefs.

The federal government should also provide funding for training, equipping and staffing of fire department special operations, including hazardous materials, technical rescue and terrorist/weapons of mass destruction response.

The subject of problem-focused research activities, supported by the federal government, has been addressed elsewhere in these findings and recommendations. A critical component of such research should be the funding of additional research in fire fighter protective clothing and equipment. Appropriate government agencies should also provide consistent certification, testing, field research and when necessary, product recall of all fire fighter protective clothing and equipment.

FINDING #10

Emergency Medical Services

As discussed earlier, today's fire services confront the full range of hazards and risks for America's communities. Primary emergency medical response to incidents that require rescue operations has become a dominant role of the fire services. In the last ten years, Emergency Medical Services (EMS), ranging from primary response to advanced life support, have grown to occupy a particularly unique and prominent position - virtually the "gate-keepers" of the health and medical service when trauma or emergency are involved. Emergency Medical Technicians (EMT's) and Paramedics have a greater level of training than ever before and are as much a part of the health care environment as they are of the firefighter environment.

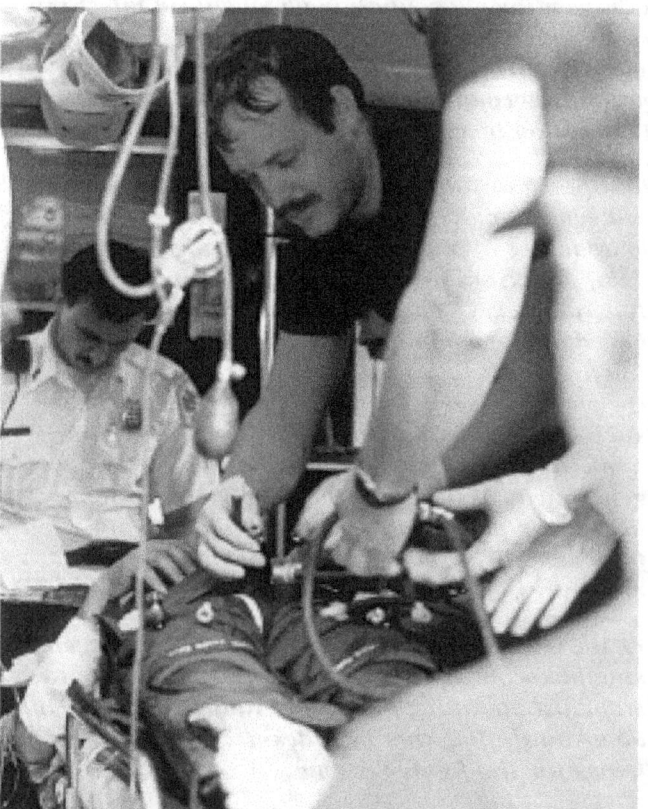

Not all fire departments and their employees have welcomed the larger role of EMS. Emergency medical response often requires significant financial resources for each emergency call, regardless of the number and nature of medical emergencies that necessitated the call. In addition, the personal and interpersonal skills needed for EMS often differ sharply from those needed to suppress a structural fire. This aspect has hindered the professional development of many in the fire services, both career and volunteer, with respect to their EMS responsibilities.

The budgets in many fire departments favor fire suppression at the expense of EMS, particularly in the area of training. EMS systems often provide inconsistent response because of this insufficient training as well as insufficient staffing. The result too often is a failure to deploy and arrive at a medical emergency within medically acceptable response times.

Federal support currently provided to the fire services' EMS component is inadequate and EMS suffers from a lack of broad programmatic support and close working relationships with the health care and health insurance industries.

Recommendations:

Support for EMS should include advocacy, improved training and equipment, research and data improvements. Strategies should be implemented that improve the practical equality of EMS within the fire service. Simply put, EMS should be adequately funded and staffed. Achieving this adequacy is the joint responsibility of government and the health care system. Emergency medical service delivery should be consistent with medically acceptable response times through the deployment of sufficient numbers of trained personnel. Fire departments should be accountable for activities conducted at the defined incident location as well as for other emergency location safety, including the provision of adequate personnel prior to the commencement of operations.

Each fire department, volunteer and career alike, should assess the EMS training needs of its current staffing. Training programs that treat career and volunteer members differently should be eliminated. Training policies that allow senior members to avoid enhanced training when newer members must obtain it should also be eliminated.

FEMA should review the collective support provided by the federal sector to the EMS activity of communities' fire departments and, based on a needs assessment, determine whether that support can be revised in order to enhance the EMS capability of these departments.

FEMA should facilitate the development of a working partnership among the health care industry, the health insurance industry, and the fire services with the goal of enhancing the provision of emergency medical services to the public and improving the efficiency and effectiveness of the health service industry.

FINDING #11

Diversity

Today's fire service has a diverse membership. Through the initiation of public policies intended to enhance the diversity of community fire departments, the face of the fire service in most metropolitan areas has changed significantly since the publication of *America Burning* in 1973. The fire service today is more inclusive of minorities and women. There has been a giant leap forward from the era in which minority representation was limited to certain stations in certain areas and there were no women firefighters. However, although the overall membership of the fire service has become more diverse, there are still a number of fire departments in which diversity does not exist or where there are barriers that limit either the upward or lateral mobility of minorities and women, irrespective of merit. There is still much to be done in building diversity into the service's organizational structure and the agenda of the emergency services.

Surveys clearly show that the most trusted societal element of today's villages, towns, cities and counties are the members of the fire services. Fire service departments and organizations have the closest personal relationship with the neighborhoods in which they operate and should be extremely reflective of our communities. Much of the strength of the fire and emergency services derives from their acceptance by the communities and neighborhoods they serve. This strength is enhanced to the extent that the fire services reflect the make-up of the community they serve.

COURTESY PRINCE GEORGE'S COUNTY FIRE DEPT.

Recommendations:

In order to improve fairness and diversity within the fire services, there should be a commitment to alter traditional attitudes with respect to the activities that are most important to the fire services. There should be recognition for those leaders and departments that effectively put an end to those traditions that limit evolution toward a diverse fire and emergency services organization.

Such leaders should establish policies and practices that improve the lateral and upward mobility of all, based on merit, and should enhance the connection of the firehouses to their neighborhoods. Both firefighters and their organizational management representatives should address the issues of fairness to all employees within their organizations.

The conduct of activities and initiatives that are intended to diminish improper imbalances with respect to diversity within a fire department should also be directed outside of the department, toward the community and the neighborhood. Fire plans and general response plans that are developed for the community should anticipate the additional concerns and challenges that occur in diverse communities, such as communication challenges, requirements for faith-related practices, societal habits and mores, and safety requirements. In addition, diversity should be considered in the conduct of prevention and preparedness activities, not only to anticipate the concerns that will arise in the response environment, but also to take advantage of the diversity achieved within the department and enhance the effectiveness of prevention and preparedness messages.

FINDING #12

Burn Injuries and Care

The trauma caused by burn injuries to civilians and firefighters is well understood within the medical community. Prompt treatment of a burn victim at a burn center as opposed to most hospitals usually is the difference between life and death. Burn survival has improved significantly over the past thirty years. On a yearly basis, deaths, once the victim has been placed into the burn care system, have decreased from around 4000 to 1000. Today, over 100 centers provide burn care, with 25 of them being full service burn treatment, research and rehabilitation. In comparison, only 12 facilities then were capable of offering a full spectrum of burn care treatment when the original America Burning report was issued.

Unfortunately, the current trend in burn care treatment, research and rehabilitation services indicates the progress has stagnated and in many respects regressed in medical research and available treatments. The Commission heard testimony that economics are discouraging many hospitals from continuing their emphasis on burn treatment. The high cost to hospitals of burn treatment and limitations on reimbursement under many existing insurance policies is currently driving down the quality and quantity of burn treatment facilities. However shortsighted this approach may be, it still exists.

Moreover, the United States has not given priority to either the broad distribution of information or the development of the technology to treat burn victims in a comprehensive manner. The federal government has actually decreased its financial investment in burn injury issues and fewer federal burn facilities exist today than did in 1973.

Recommendations:

Prompt and comprehensive care for the burn victim is essential, benefiting not only the victim, but also society as a whole. This care should not be limited to the physical needs of the victim alone, but should be expanded to consider the mental and emotional needs of the victim and his or her family, friends, and often times, co-workers.

FEMA and the United States Fire Administration should build partnerships that will support both the prevention and care giving and expand the capability to manage all aspects of burn-related issues.

With regard to treatment, these partnerships:

★   Should include advocating within the health industry the needs of victims. This advocacy should impress on insurers the benefits of immediate and comprehensive treatment as contrasted with the alternative costs of delays caused by inadequate insurance coverage; and

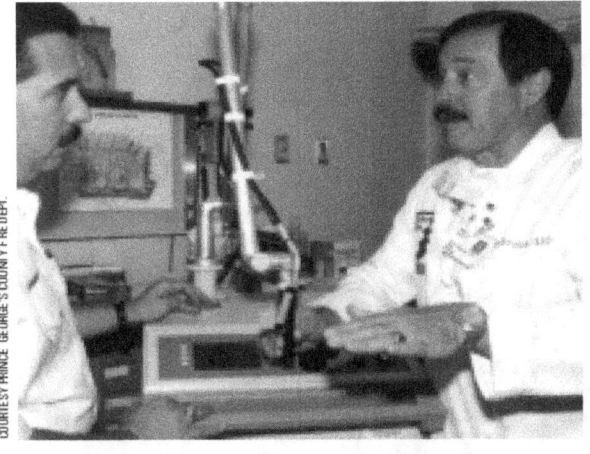

★   Should lead to the maintenance of training centers, the development of programs to recruit and retain burn physicians and nurses, and an increase in federal research such as that once provided by Brook Army Medical Center.

AMERICA BURNING
Recommissioned

Appendices

FIRE CONTROL AND PREVENTION IN THE U.S.: A FEDERAL PERSPECTIVE

With the enactment of Public Law 93-498, the "Federal Fire Prevention and Control Act of 1974," Congress established a Federal institutional focus for the Nation's fire problem. This institution, originally called the National Fire Prevention and Control Administration (NFPCA), is now known as the United States Fire Administration (USFA). Today, it has become a widely utilized resource by fire departments, fire safety professionals, and injury prevention advocates all across America.

Events Behind The Creation of USFA

A number of studies and reports were driving forces in establishing the need for an agency such as USFA. A 1966 report, the *Wingspread Conference on Fire Service Administration, Education and Research: Statements of National Significance to the Fire Problem in the United States,* challenged the traditional concept of fire protection being strictly a local responsibility. As a result of this report and of developments in the public policy climate that signaled a willingness to explore ways to provide federal support to state and local public safety issues, national fire and allied groups began to push for Federal leadership, advocacy, and assistance to help state and local governments meet their fire protection and public safety responsibilities.

Although a Presidential Commission was authorized by Federal legislation to study the problem of fire in America via the "Fire Research and Safety Act of 1968," little or no action was undertaken until national fire service organizations agreed to work together to find a way to move the effort forward. Meeting in Williamsburg, Virginia, in the summer of 1970, national fire services groups working together identified approaches to bring pressure on the White House to actually begin a serious Federal study of the national fire problem and the role of the Federal government.

In 1971, President Richard M. Nixon appointed 24 individuals to the National Commission on Fire Prevention and Control, which undertook a challenging two-year study focused on fire in America. The Commission was made up of a broad spectrum of private and public sector representatives including members of the Congress, fire industry leaders, heads of fire services groups, and burn prevention advocates, as well as labor and management officials. After extensive work, the group issued their landmark document, *America Burning,* in May of 1973.

This seminal report detailed the Nation's unacceptably high losses from fire and identified the appropriate role for the Federal government in the effort to substantially reduce these losses. The kind of national leadership and advocacy called for in the report continue to be centerpieces of USFA activity. The report enjoyed nationwide acceptance and became the catalyst for Public Law 93-498, the "Federal Fire Prevention and Control Act of 1974."

At the time the legislation was enacted, there was universal agreement on the need for an effective Federal entity to provide national leadership in dealing with the nation's fire problem. In addition to the need for Federal advocacy to support the work of operational fire personnel, the key role of those involved in improving building codes and standards and in making technological advances with application to fire safety was also universally recognized. Mechanisms for improving linkages between public policy makers and the private sector were viewed as critical lynchpins in ensuring effective advocacy for the fire service and its allied professionals.

The passage of P.L. 93-498 on October 29, 1974 marked an historic event that would have a major impact on the quality of fire prevention and control in this country forever after. The Act, passed by the Congress and signed by the President, officially led to the creation of the NFPCA (now USFA), which was originally established within the U.S. Department of Commerce. For over 25 years, the agency has provided the principal Federal focus on fire-related issues. The primary reason for the initial placement of the Federal Fire

Programs within the Commerce Department was that this Department also housed the Center for Fire Research within its National Bureau of Standards. Today, this entity is known as the Building and Fire Research Laboratory of the National Institute of Standards and Technology.

The USFA in FEMA

Five years after the passage of USFA's enabling legislation, on April 1, 1979, the USFA and its related training agency, the National Fire Academy (NFA), were placed into the newly created Federal Emergency Management Agency (FEMA). FEMA experienced many of the typical challenges that can confront an organization compiled from other existing organizations. FEMA was created, in part, to respond to the concerns of State and local officials about the fragmentation of Federal disaster relief programs. Many in the fire services questioned the placement of the agency focused on reducing fire loss into an agency that was to be focused on large-scale catastrophic disasters - not the type of emergencies the fire service handles every day. Others, however, believed that the inclusion of the USFA within FEMA was a logical step when one considered the role of the first responder community in dealing with disasters of all kinds.

An early decision (1980) by FEMA to separate the USFA from the National Fire Academy underscored a fear that efforts were being made in FEMA to dismantle Federal fire programs. During FEMA's first ten years, fire services frequently questioned what was described as a lack of support and advocacy from FEMA. An unintended and unfortunate side effect of the tension was a perception that the fire services were unwilling to work together to support improvements and initiatives that were framed to link fire and emergency management groups. In January 1991, FEMA reunited the USFA and NFA program elements under USFA administration. In 1998, under Director James Lee Witt, FEMA articulated two priorities for the Agency with respect to the fire services in America:

1. Acknowledgement and advancement of the fire services by FEMA and support for them as the critical component of the first responder community, and

2. Emphasis on the need for collaboration and mission integration between the fire services and emergency management agencies.

In the summer of 1998, the Director formed a Blue Ribbon Panel in order to examine concerns that representatives of varying aspects of the fire services had expressed about FEMA and the USFA and in order to assess the future role of the USFA. The Blue Ribbon Panel was asked to consider the agency's role in light of the many changes impacting the fire service and its evolving needs. The report of the Blue Ribbon Panel and its companion Action Plan document, which the USFA had developed in order to carry through on Blue Ribbon Panel's recommendations, provide the blueprint for principally inward-looking improvements for FEMA and USFA. However, the broader context of the nation's current fire problem still needed to be discussed and addressed. In the summer of 1999, the FEMA Director, acting on a recommendation of the Blue Ribbon Panel, recommissioned a panel to continue and update the work of the National Commission on Fire Prevention and Control, which produced the 1973 *America Burning* report.

STATUS OF RECOMMENDATIONS OF AMERICA BURNING

(those directed toward "non-USFA" institutions)

The America Burning report of 1973 included 90 recommendations. As noted elsewhere in this report, the 1973 report was seminal and comprehensive in its identification of the issues confronting the fire services community. In the opinion of the members of America Burning, Recommissioned, an assessment of the status of these recommendations is important to understand those issues in the present day, their practical context, and the findings and recommendations that have been formulated to advance the resolution of those issues.

This chapter describes the status of the recommendations that the 1973 report had indicated, or that America Burning, Recommissioned had interpreted, to be within the control of agencies other than the U.S. Fire Administration. There have been organizational changes to the federal sector since the 1973 report; therefore, each recommendation identifies the agency that supplied an update on the status of work and the date on which this information was supplied. Recommendations within the responsibility of the USFA to address are not included here since the report of the Blue Ribbon Panel convened in 1998 has already assessed the status of implementation of those 1973 recommendations that were within the direct control of USFA.

Recommendation #5

The National Institutes of Health should greatly augment their sponsorship of research on burns and burn treatment.

Recommendation #6

The National Institutes of Health should administer and support a systematic program of research concerning smoke inhalation injuries.

Agency: **Department of Health and Human Services – Public Health Service – National Institute of Health (NIH)**
Date: **February 25, 2000**

★ The National Institute of General Medical Science (NIGMS) has a discrete and diverse program in Trauma and Burn Injury.

★ It should be noted that burn-related research is considered a subset of trauma, and thus is not tracked or administered separately from traumatic injury as a whole.

★ The Trauma and Burn Injury Program at NIGMS supports research project grant awards to investigators from all disciplines. The program is designed to accommodate research from the molecular level up to and including outcome studies on injured patients, and ranging from the immediate post-injury response to resolution of the injury including studies of any complications.

★ Dramatic improvements have taken place over the past 20 to 30 years in the United States, particularly in the major burn centers. The recent advances in the care of burn victims may be viewed as both a direct and indirect consequence of support from NIGMS in particular and NIH in general. Additionally, numerous advances in the fundamental knowledge of the pathophysiology of burn injury have been made that have not yet been extended to clinical practice. Many questions and problems remain to be addressed and solved.

★ It should be emphasized that NIGMS is not the sole source of research support for burn injury. Many other Institutes of NIH also fund a variety of research programs, including the National Institute on Alcohol Abuse and Alcoholism; the National Institute of Allergy and Infectious Diseases; the National Institute of Arthritis and Musculoskeletal and Skin Diseases; the National Institute of Child Health and Human Development; the National Institute of Mental Health; and the National Heart, Lung, and Blood Institute.

★ Brief descriptions of these Institutes' relevant research projects are as follows:

   National Institute of Alcohol Abuse and Alcoholism (NIAAA): The NIAAA currently supports two projects that investigate the effects of alcohol on trauma, including burn injury.

   National Institute of Arthritis and Musculoskeletal and Skin Diseases (NIAMS): The NIAMS has a significant portfolio of projects on wound healing, including healing that follows burn wounds.

   National Institute of Child Health and Human Development (NICHD): The NICHD is pursuing novel interventions to burn injury management that fall into two basic categories: pain management and tissue regeneration.

   National Institute of Mental Health (NIMH): The research portfolio of NIMH in the area of traumatic stress includes studies focusing on the biological, psychological, and social factors that influence the mental health consequences of trauma, including fires and burn injuries.

   National Heart, Lung, and Blood Institute (NHLBI): The NHLBI does not have a specific program on burns and non-tobacco smoke inhalation injury, although the Institute has an active interest in research related to smoke inhalation injuries.

Recommendation #15

The commission urges the federal research agencies, for example, the National Science Foundation and the National Institute of Standards and Technology, to sponsor research appropriate to their respective missions within the areas of productivity of fire departments, causes of fire fighter injuries, effectiveness of fire prevention efforts and the skills required to perform various fire department functions.

Agency: Department of Commerce -
             National Institute of Standards and Technology (NIST)
Date: February 14, 2000

★ NIST developed fire safety educational booklets for the USFA and a new series of fact sheets on fire safety in manufactured homes.

★ The Building and Fire Research Laboratory (BFRL) developed one of the first aids to fire investigations. Through measurements of material burning characteristics in laboratory and large-scale experiments and through application of fire models developed at BFRL, the conditions that led to firefighter injuries or deaths in a number of actual incidences have been analyzed. The results have suggested procedures to prevent recurrences.

★ BFRL is investigating the thermal performance of the fire fighter protective clothing under relatively low heat flux exposures. The combination of data and the understanding embodied in user-friendly software can aid fire fighter equipment manufacturers in the design of cost-effective, higher performance protective ensembles. This technology can also be cast in a slightly different form to train fire fighters about the limitations of their protective equipment.

★ BFRL is working with the detection and alarm industry to enable the display and transfer of vital building information during fire emergencies.

★ BFRL is developing a standard icon-based system for use in the visual display of information in advanced fire alarm panels and probable fire service hand-held decision aid devices of the future.

★ BFRL is quantifying the marks left by experiments that simulate arson and natural fire events to develop the technical basis for reliable fire investigation tools.

Recommendation #24

The commission urges the National Science Foundation, in its Experimental Research and Development Incentives Program, and the National Institute of Standards and Technology, in its Experimental Technology Incentives Program, to give high priority to the needs of the fire service.

Agency: **Department of Commerce - National Institute of Standards and Technology**
Date: **February 14, 2000**

★ The National Bureau of Standards (NBS) and NIST grants program has supported a number of areas addressing the needs of the fire service, including:

- Fire fighter safety
- Fire suppression with water
- Life safety
- Detection, and
- General topics.

Recommendation #30

The new Consumer Product Safety Commission (CPSC) should give a high priority to the combustion hazards of materials in their end use.

Agency: **United States Consumer Product Safety Commission (CPSC)**
Date: **March 3, 2000**

★ This recommendation is incorporated into the Commission's strategic goal of saving lives and keeping families safe from fire-related causes.

★ CPSC special studies provide more detailed information on the involvement of consumer products in fire injuries and death than other studies.

★ The CPSC analyzes this information to identify fire scenarios that require urgent attention, then develops and implement strategies to reduce the problem and make the best use of resources.

★ CPSC's efforts to reduce fire hazards include working with industry to develop and improve voluntary safety standards and to promulgate mandatory safety standards.

★ Ongoing compliance activities identify and correct product defects that pose fire hazards, monitor conformance with voluntary fire safety standards and codes, and ensure compliance with our mandatory regulations.

★ The CPSC also conducts public information campaigns and works in partnerships with other federal agencies and interested groups on all aspects of fire safety involving consumer products. These include:

- CPSC fire safety codes

- Clothing flammability

- Household furnishings/materials

- Devices for fire/gas safety

- Residential electrical wiring

- Household heating and electrical appliances, and

- Ignition sources.

Recommendation #31

The present fuel load study sponsored by the General Services Administration (GSA) and conducted by the National Institute of Standards and Technology (NIST) should be expanded to update the technical study of occupancy fire loads.

Agency: Department of Commerce - National Institute of Standards and Technology
Date: February 14, 2000

★ The GSA industrial fire load study was extended in 1977 by the National Bureau of Standards (NBS) to fire loads in residential occupancies, under the sponsorship of the U.S. Department of Housing and Urban Development (HUD).

★ NBS/NIST researchers developed oxygen consumption calorimetry, which on a small scale lead to the invention of the cone calorimeter, and on a larger scale to the furniture calorimeter.

Agency: U. S. General Services Administration – Public Building Service
Date: February 28, 2000

★ A Survey of Fuel Loads in Contemporary Office Buildings was completed in November of 1995. Two methods for determining movable fuel load were utilized in this study. Direct weighing techniques were utilized in both methods.

★ Surveys were conducted in buildings at the University of Maryland College Park and at the General Services Administration (GSA) Headquarters Building in Washington, D.C. Statistical results are presented for the two survey methods, each office type, and each material category.

★ The results of the study present the impact of open plan designs on the fuel load and also present the partition and computer accessory fuel load contributions.

Recommendation #32

The Consumer Product Safety Commission should give high priority to flammability standards for fabrics.

Agency: **United States Consumer Product Safety Commission (CPSC)**
Date: **March 3, 2000**

★ This recommendation is incorporated into the Commission's strategic goal of saving lives and keeping families safe from fire-related causes.

★ CPSC special studies provide more detailed information on the involvement of consumer products in fire injuries and death than other studies. The CPSC analyzes this information to identify fire scenarios that require urgent attention. It then develops and implements strategies to reduce the problem and make the best use of resources.

★ CPSC's efforts to reduce fire hazards include working with industry to develop and improve voluntary safety standards and promulgating mandatory safety standards.

★ Ongoing compliance activities identify and correct product defects that pose fire hazards, monitor conformance with voluntary fire safety standards and codes, and ensure compliance with our mandatory regulations.

★ The CPSC also conducts public information campaigns and works in partnerships with other federal agencies and interested groups on all aspects of fire safety involving consumer products. Examples include:

    CPSC fire safety codes,

    Clothing flammability

    Household furnishings/materials

    Devices for fire/gas safety

    Residential electrical wiring

    Household heating and electrical appliances, and

    Ignition sources.

Recommendation #34

The Department of Commerce should be funded to provide grants for studies of the dynamics of combustion and the means of its control.

Agency: **Department of Commerce - National Institute of Standards and Technology**
Date: **February 14, 2000**

★ Since 1978, NBS/NIST has administered the extramural fire research program at an annual level between $1.3M and $2.0M. Some 20 to 30 grants have been awarded each year, with duration up to three years.

★ Taken together with the internal fire research programs of the BFRL, the combined research output of NIST is the largest source of new knowledge of fire behavior in the world. Findings are documented in thousands of journal articles and reports, and the results from this research have revolutionized our approach to predicting fire risk and mitigating fire losses.

Recommendation #35

The National Institute of Standards and Technology and the National Institutes of Health should cooperatively devise and implement a set of research objectives designed to provide combustion standards for materials to protect human life.

Agency: **Department of Commerce - National Institute of Standards and Technology**
Date: **February 14, 2000**

★ The Center for Fire Research introduced the notion of fire scenarios and analyzed fire incident data to determine the scenarios that most frequently led to fire death.

★ This led to the following test methods which have played a major role in the dramatic reductions in fire deaths in homes and other buildings since then:

  ⋆ Children's sleepwear test method

  ⋆ Floor covering and attic insulation radiant panel test method

  ⋆ Cigarette/smoldering ignition test

  ⋆ Cone, furniture, and room fire calorimeters

  ⋆ Mattress flammability test method NIST

  ⋆ Smoke toxicity test method

  ⋆ Criteria for fire-safe installation of fireplaces and fireplace inserts, and the

  ⋆ Cigarette ignition propensity test method.

★ NIST scientists are currently at the cutting edge of flame retardant research, seeking to ensure that the environmental push to eliminate traditional halogenated retardants from plastics will not leave the public unprotected.

★ NIST research on a range of issues including interior finish, electrical, smoke alarms, and sprinklers, led to the improved HUD standard.

★ NIST pioneered the application of fire models as an investigative tool.

Recommendation #36

The commission urges the National Institute of Standards and Technology to assess current progress in fire research and define the areas in need of additional investigation. Further, the institute should recommend a program for translating research results into a systematic body of engineering principles and, ultimately, into guidelines useful to code writers and building designers.

Agency: **Department of Commerce - National Institute of Standards and Technology**
Date: **February 14, 2000**

★ NIST has worked with the Society of Fire Protection Engineers (SFPE) to transfer advances in the understanding of fire and the tools for fire hazard prediction to fire protection engineers and allied professionals.

★ NIST research contributes to the development of quantitative methods for fire hazard and risk assessment and performance-based fire codes.

★ The enabling technology for modern fire safety engineering is the development of fire modeling, and NIST is recognized as the founder and world leader in this area. What brought modeling to the point of being a common engineering tool were two unforeseen developments – the age of the ubiquitous personal computer and the publication by BFRL in 1988 of HAZARD I, still the world's first and only comprehensive fire hazard assessment software tool.

★ There is a worldwide move toward performance-based codes that promises to reduce the economic burdens of providing fire safe buildings without reducing safety.

★ Field models, such as BFRL's high spatial and temporal resolution fire model. This model brings the technology of computational fluid dynamics to fire safety engineers for application on personal computers, where most of the world's engineering is done.

Agency: Society of Fire Protection Engineers
Date: January 20, 2000

★ The final report on developing a research agenda, which will be published shortly, identifies research most needed by the fire protection engineering community to improve life safety, reduce fire-related costs, and improve environmental protection.

★ The society has developed design guides that transfer research into practice in the areas of thermal radiation from pool fires and skin burns from thermal radiation. Design guides are under development in several other areas, including human behavior in fire, fires designed to simulate and predict burn effects, and room origin fire hazards.

Recommendation #37

The National Institute of Standards and Technology, in cooperation with the National Fire Protection Association and other appropriate organizations, should support research to develop guidelines for a systems approach to fire safety in all types of buildings.

Agency: Department of Commerce - National Institute of Standards and Technology
Date: February 14, 2000

★ The availability of modeling as a means to quantify the fire performance of buildings and facilities, as noted in the previous paragraphs, has provided the critical link that allows the meaningful application of a systems approach to fire safety, or as we prefer, engineered fire safety.

Agency: Society of Fire Protection Engineers
Date: January 20, 2000

★ The Society of Fire Protection Engineers (SFPE) recently completed The SFPE Engineering Guide to Fire Protection Analysis and Design of Buildings. This guide describes a method for considering the contributions of all fire safety systems in a building towards fire safety. As such, it is the first guide published in the United States that outlines a systems approach to fire safety.

Recommendation #39

The commission urges the Consumer Product Safety Commission to give high priority to matches, cigarettes, heating appliances and other consumer products that are significant sources of burn injuries, particularly products for which industry standards fail to give adequate protection.

Agency: **United States Consumer Product Safety Commission (CPSC)**
Date: **March 3, 2000**

★ This recommendation is incorporated into the Commission's strategic goal of saving lives and keeping families safe from fire-related causes.

★ CPSC special studies provide more detailed information on the involvement of consumer products in fire injuries and death than other studies. The CPSC analyzes this information to identify fire scenarios that require urgent attention and then develops and implement strategies to reduce the problem and make the best use of resources.

★ CPSC's efforts to reduce fire hazards include working with industry to develop and improve voluntary safety standards and to promulgate mandatory safety standards.

★ Ongoing compliance activities identify and correct product defects that pose fire hazards, monitor conformance with voluntary fire safety standards and codes, and ensure compliance with our mandatory regulations.

★ The CPSC also conducts public information campaigns and works in partnerships with other federal agencies and interested groups on all aspects of fire safety involving consumer products. Examples include:

  - CPSC fire safety codes

  - Clothing flammability

  - Household furnishings/materials

  - Devices for fire/gas safety

  - Residential electrical wiring

  - Household heating and electrical appliances, and

  - Ignition sources.

Recommendation #41

The commission urges the Society of Fire Protection Engineers to draft model courses for architects and engineers in the field of fire protection engineering.

Agency: **Society of Fire Protection Engineers**
Date: **January 20, 2000**

★ The Society of Fire Protection Engineers has developed and offered a number of courses on fire protection engineering. These include:

  - Introduction to Performance-Based Design for the Authority Having Jurisdiction

  - Smoke Management for Atria and Other Large Spaces

    Introduction to Sprinkler Design for Engineers

    Introduction to Computer Fire Modeling, and

    Advanced Fire Modeling.

★ The Society of Fire Protection Engineers has recently entered into agreements to develop additional courses on performance-based design for the enforcement community.

Recommendation #50

The Department of Treasury should establish adequate fire regulations, suitably enforced, for the transportation, storage and transfer of hazardous materials in international conference.

Agency: **United States Department of Treasury**
Date: **March 17, 2000**

★ Since the issuance of the 1973 Report, programmatic responsibility for these issues was transferred from the Department of the Treasury to the Department of Transportation. Therefore, it is appropriate that the Department of Transportation should address any review of the above recommendation.

Recommendation #57

The Department of Agriculture's assistance to community fire protection facilities projects should be contingent upon an approved master plan for fire protection for local jurisdictions.

Agency: **United States Department of Agriculture**
Date: **May 26, 2000**

★ The USDA has developed has developed a group of loans and grants available to qualifying communities based on their needs as determined through community planning. This includes the construction of fire and emergency fatalities.

★ Opportunities provided by the USDA include:

    The Community Facilities Direct Loan Program for nonprofit and public entities and construction of essential facilities

    The Community Facilities Guaranteed Loan Program, which provides 90 percent loans from a private lender for essential facilities, and

    The Community Facilities Grant Program, used to fund projects under special initiatives, such as Native American community development activities.

Recommendation #58

The proposed U.S. Fire Administration should join with the U.S. Forest Service in exploring means to make fire safety education for forest and grassland protection more effective.

Agency: United States Department of Agriculture – Forest Service
Date: February 22, 2000

★ In this area of fire safety, the Forest Service, U.S. Fire Administration, National Fire Protection Association, National Association of State Foresters, and the U.S. Department of the Interior have developed the national FIREWISE communities program.

Recommendation #59

The Council of State Governments should develop model state laws relating to fire protection in forests and grasslands.

Agency: The Council of State Governments
Date: January 25, 2000

★ The 1975 Suggested State Legislation (SSL) contains "An Act to Facilitate Exchange of Fire Control Manpower and Equipment." This resulted from recommendation #59.

★ The states and federal government have organized four Forest Fire Company organizations (the Mid-Atlantic, Southeastern, Northeastern and South-Central) involving 26 states. These companies have become the forest fire protection compacts that are in force today.

Recommendation #61

The U.S. Forest Service should develop the methodology to make possible nation-wide forecasting of fuel build-up as a guide to priorities in wildland management.

Agency: United States Department of Agriculture – Forest Service
Date: February 22, 2000

In the area of fuel build-up forecasting, two significant steps are complete.

★ The first is the "Fire Regimes for Fuels Management and Fire Use Project," begun in 1999.

★ The second project is called "Ecosystems at Risk."

Recommendation #62

The commission supports the development of a National Fire Weather Service in the National Oceanic and Atmospheric Administration and urges its acceleration.

Agency: U. S. Department of Commerce -
           National Oceanic and Atmospheric Administration - National Weather Service
Date: February 11, 2000

★ The National Weather Service (NWS) has undergone an extensive modernization and restructuring effort.

★ The NWS has developed national standards for fire weather support.

★ The NWS has expanded its capability to respond to wildfire and other hazardous incidents.

Recommendation #76

The National Fire Protection Association and American National Standards Institute should jointly review the Standard for Mobile Homes and seek to strengthen it, particularly in such areas as interior finish materials and fire detection.

Agency: National Fire Protection Association (NFPA)
Date: February 3, 2000

★   The U.S. Department of Housing and Urban Development (HUD) upgraded all interior finish materials' fire safety requirements in 1976, as part of the Mobil Home Construction and Safety Act of 1974.

★   Fire detection requirements now need to be modernized by HUD. On February 1, 1999, NFPA recommended changes to the smoke alarm standards. Action on these recommendations is pending, although we have been assured HUD intends to enter into rulemaking to upgrade the smoke alarm standards.

Recommendation #77

All political jurisdictions should require compliance with the NFPA/ANSI standard for mobile homes, together with additional requirements for early-warning fire detectors and improved fire resistance of materials.

Agency: National Fire Protection Association
Date: February 3, 2000

★   With passage of Public Law 93-383 with implementation starting in 1976, NFPA/ANSI 501B became the preemptive federal standard for the entire manufactured housing industry under the jurisdiction of the U.S. Department of Housing and Urban Development (HUD). Fire safety has seen significant improvement as a result of the federal preemptive codification of the NFPA/ANSI 501B standard.

★   Because of the 1974 public law codifying the NFPA/ANSI 501B standard as a preemptive federal regulation, all states/political jurisdiction have been required to comply with the NFPA/ANSI consensus standard (95% of the standard was codified into the HUD regulation).

Recommendation #84

The National Institute of Standards and Technology should develop standards for the flammability of fabric materials commonly used in nursing homes, with a view to providing the highest level of fire resistance compatible with the state-of-the-art and reasonable costs.

Agency: Department of Commerce - National Institute of Standards and Technology
Date: February 14, 2000

★   Many of the actions taken in response to recommendations 34 an 35 apply to the problem of fabric flammability experienced in nursing homes. The test methods developed for floor coverings, mattresses, and cigarette ignition are of direct benefit to nursing homes.

★ Research has been conducted on active fire protection systems to enhance life safety in institutional settings. This research, coupled with studies on controlling smoke generation and spread, assisted in addressing this recommendation.

No information was received from the following organizations as of May 3, 2000:

National Science Foundation, Item 15

Joint Council of National Fire Service Organizations, Item 28

American Institutes of Architects, Item 40

Department of Transportation, Items 51 and 53

Urban Mass Transportation Administration (DOT), Item 55, or

IRS, Item 73.

Members of **AMERICA BURNING:** *Recommissioned*

Front Row, (kneeling): Al Whitehead; Carol Shelton; Ron Siarnicki; Jane Smith; Herman Brice; Bill Peterson; 2nd Row: Rich Marinucci; Dean Ockerbloom; George Bernstein (Chairman); Vina Drennan; Cynthia Wilk; FEMA Director James L. Witt; Bill Pessemier; Ken Burris (ex officio); Last Row: Jack McElfish; John Best; Mike Bell; Louis Fiore; Melvin Stark; Glenn Corbett; John Buckman [not pictured: Ronald Coleman; John Eversole; Lamont Ewell; Arthur Glatfelter]

GEORGE K. BERNSTEIN (CHAIRMAN)

George K. Bernstein (Chairman), the first Federal Insurance Administrator, a former First Deputy Superintendent and General Counsel of the NY State Insurance Department and a NY State Assistant Attorney General, is currently an attorney in Washington, D.C. and New York City. He has extensive career experience representing State insurance departments, organizations and corporations on a full range of insurance matters. He often testifies in courts and before State legislatures and the Congress as an expert witness. He is also the author of numerous articles and studies and has lectured on insurance, natural hazards management and policy, and in other areas.

Mr. Bernstein is well known for his seminal work in developing and implementing, as the first Federal Insurance Administrator, the National Flood Insurance Program, which is the Federal government's first comprehensive hazards management program that includes hazard identification, hazard reduction and financial incentives within itself. Mr. Bernstein also administered Federal FAIR Plan and Riot Reinsurance Program, and the Federal Crime Insurance Program.

Mr. Bernstein has served on national and international committees and boards that have been focused on a full range of insurance issues. He was U.S. delegate to the NATO Conference on Flood Insurance. He also served on and advised the Cost of Living Council and the Price Commission. He was a member of the President's Interdepartmental Committees on All-Risk and Disaster Insurance and the National Academy of Sciences Committee on Medical Malpractice. He has consulted to numerous organizations, including the Inter-American Development Bank and the Federal Retirement Investment Board. He has continued to provide invaluable formal and informal support and advice to FEMA Directors on hazards management issues. He chaired from 1986 to 1990 the Federal Expert Review Committee of the National Earthquake Hazards Reduction Program, the NEHRP Advisory Committee from 1991 to 1993 and the 1998 Federal Ad Hoc Panel for the Development of a National Pre-disaster Mitigation Plan.

Mr. Bernstein received his BA and LLB from Cornell University and has been admitted to the New York Bar, the U.S. Court of Appeals (2nd Circuit), the U.S. Supreme Court and the District of Columbia Bar. Included among his awards are the HUD Distinguished Service Award and the Torch of Liberty Award from the Anti-Defamation League of the B'nai B'rith.

MICHAEL P. BELL

Michael P. Bell has served as the Director/Chief of the City of Toledo Department of Fire and Rescue Operations since 1990. He served in a number of positions prior to this appointment including Water Rescue Diver, Fire Recruiter, Paramedic, Paramedic Shift Supervisor, Training Officer (on State and National levels), and Training Captain. He is the first black to be appointed to the position of Chief in the history of the fire service in the City of Toledo as well as being the youngest fire chief running a metro fire department in the United States at age 35.

Chief Bell holds a Bachelor of Education Degree from the University of Toledo. He has served on the boards of the Cystic Fibrosis Association, the American Red Cross, the Intercity Youth Movement, and United Way of Greater Toledo. Chief Bell was selected to carry the Olympic Torch in June 1996 through the City of Toledo as a community leader.

CHIEF JOHN M. BEST

Chief John M. Best is Manager of Emergency Services for the Reedy Creek Improvement District, providing code administration, fire, emergency medical and related services to Walt Disney World, Florida. He has provided fire protection services through the International Association of Fire Chiefs (IAFC) Operation Life Safety and the National Fire Protection Association (NFPA) to more than fifteen major cities nationwide relating to model codes and standards, fire protection systems and the adoption of life and fire safety legislation. He is currently appointed to the NFPA Technical Committee on Fire Service Training and serves on the Editorial Review Board of the International Society of Fire Services Instructors.

Mr. Best retired as Deputy Fire/Rescue Chief, having served twenty-eight years in all administrative and operational levels of the Montgomery County, Maryland, Department of Fire and Rescue Services. He holds a Master of Arts degree in Organizational Management with degrees in Business Administration and Fire Science from the University of Phoenix. He has served as adjunct faculty for Montgomery College (MD) and currently serves on the Fire Science Advisory Board of Valencia College (FL). Mr. Best has received three Governor's Citations from the Governor of the State of Maryland and numerous national, state and local awards and commendations recognizing his dedication, performance and accomplishments.

CHIEF HERMAN BRICE

Chief Herman Brice has served as Fire Rescue Administrator at the Palm Beach County Fire-Rescue Department in the State of Florida for fifteen years. He has forty-six years of fire-rescue experience including thirty-one years of service with the City of Miami Fire, Rescue & Inspection Service Department in Florida, and as Fire Chief for six years.

Chief Brice has chaired the National Fire Protection Association Board of Directors, the Metropolitan Fire Chiefs' IAFC, and the Joint Council of Fire Service Organizations of Florida. He is a past Director on the Board of the National Institute of Building Sciences and has served as President for the Florida Chiefs' Association. He is a current member of the International

Association of Fire Chiefs and the Underwriters Laboratory Fire Council (Chicago, IL) and was a member of the Florida Firefighter's Standards Council. Chief Brice was named Florida Chief of the Year in 1982 and 1998.

CHIEF JOHN BUCKMAN

Chief John Buckman has served as Fire Chief of the German Township Volunteer Fire Department since 1978, and has been Indiana Deputy State Fire Marshal since 1985. Chief Buckman was named the 2nd Vice-President of the International Association of Fire Chiefs in 1999. He has served as the International Representative of VCOS/IAFC Board of Directors since 1994, and received the VCOS Superior Leadership Award in 1997. His other activities include serving on the National Fire Academy Alumni Association Board of Directors, as NFA Resident program adjunct instructor, and as NFA Course developer of the Leadership & Administration VIP course. The Indiana Fire Instructor Association named him Instructor of the Year (1988). He has been active on the following IAFC task forces: NFPA 1200, NFPA 1851, Technology Transfer, and the Professional Designation Certification Task Force. He served as President of Southwestern Indiana Survive Alive, Inc., and serves as President of the Suburban Fire Chiefs Association.

Chief Buckman has authored 54 articles for fire service publications, and worked on the Editorial Advisory Boards for both the *National Fire Rescue Magazine* and *Fire Engineering Magazine,* and *Recruiting, Training & Maintaining Volunteer Firefighters* (3rd Ed.). He has presented at seminars and conferences in 31 states and Canada. He is currently a certified Fire Protection Specialist and Uniform Fire Code Inspector.

KENNETH O. BURRIS, JR.

Ken Burris was selected as the first Chief Operating Officer of the U.S. Fire Administration (USFA) by Federal Emergency Management Agency Director James Lee Witt in September 1999. Burris oversees the day-to-day operations of USFA, including the National Fire Academy, and serves as the primary advisor to the FEMA Director and the USFA Administrator on overall operations and management of the Fire Administration.

Before joining FEMA, Burris had served as a firefighter for more than 22 years, and had been the fire chief in Marietta, Georgia, since 1992. He is credited with applying innovative techniques and strong leadership to that department, which serves as a national model.

A native of Kansas City, Missouri, Burris holds a BS degree in Safety and Fire Protection Engineering Technology from the University of Cincinnati, where he graduated cum laude. He also holds a Masters of Public Administration at Kennesaw University.

Burris has held senior positions with the Georgia Association of Fire Chiefs and the International Association of Fire Chiefs and is a past president of the Southeastern Association of Fire Chiefs.

RON COLEMAN

Mr. Ron Coleman has served in the Fire Services for 38 years and most recently has been Chief Deputy Director, Department of Forestry and Fire Protection and the California State Fire Marshal. He has an MA in vocational education from Cal State Long Beach, a Bachelor's degree in political science from Cal State Fullerton, and an Associate of Arts degree in Fire Science from Rancho Santiago College. He has been President of the International Association of Fire Chiefs, Vice President of the International Committee for Prevention and Control of Fire, and President of the California League of Cities, Fire Chiefs Department.

Mr. Coleman's career reflects his steadily increasing level of responsibility for creating and implementing fire protection policy on the local, state and national level. Among other duties, he has been Chairman of the National Fire and Emergency Services Accreditation Task Force, the Risk Hazard and Value Evaluation Project, the Urban-Wildland Interface Code Committee, and the Orange County Emergency Medical Services Committee.

GLENN CORBETT

Glenn Corbett is an Assistant Professor of Fire Science at John Jay College of Criminal Justice in New York City, Technical Editor of Fire Engineering magazine, and a Captain in the Waldwick, New Jersey Fire Department. Additionally, he serves on the New Jersey State Fire Code Council and is the Vice President (North) of the New Jersey Society of Fire Service Instructors. He holds a Master's Degree in Fire Protection Engineering from Worcester Polytechnic Institute and a professional engineer's license from the State of Texas. Previously, he served as the Administrator of Engineering Services for the San Antonio, Texas Fire Department.

VINA DRENNAN

Vina Drennan is the widow of New York City Fire Captain, John Drennan, who died from multiple burn injuries he sustained while fighting a fire in Manhattan. For forty days, he put up a valiant struggle to live, in spite of the massive burns. His story captured the hearts of New Yorkers and developed an increased appreciation for the sacrifices that firefighters are willing to make to protect the public.

Ms. Drennan is the mother of four children, ages 19-30, and resides in Manhattan. She graduated from Wagner College, and holds a Master's degree from Richmond College. She was a New York City Public School teacher for ten years. In the past five years, Ms. Drennan has made more than 60 public appearances throughout North America advocating fire safety awareness. She was featured on NBC's "*Dateline*" in January 1995. She has served as a Fire Safety Spokesperson/Advocate since 1994. Ms. Drennan has been a Mayoral Appointee for the New York City Committee on the Status of Women since 1995. She has been a member of the FDNY Fire Foundation since 1995, the National Fallen Firefighter Foundation Family Support Network since 1994, and in 1999, joined the New York State Task Force on Fire Safety Teacher Training Curriculum. She received the St. Barnabas Burn Foundation Humanitarian Award in 1997. She has published articles in the N*ew York Daily News, Redbook Magazine, The Voice, Fire Arson Investigator,* and *Firehouse Magazine.*

CHIEF JOHN M. EVERSOLE

Chief John M. Eversole is the current Hazardous Materials Coordinator of the Chicago Fire Department. He is responsible for the Hazardous Materials HIT Team and coordinates all the Fire Departments' units that make up the Hazardous Incident Task Force. Chief Eversole has also been involved in special programs, such as the Deep Tunnel Project. He has coordinated the development of the Confined Space/Collapse Rescue operations and has been working with the U.S. Department of Defense in developing a civilian emergency response program for terrorism.

Chief Eversole is the Chairman of the Hazardous Materials Committee of the International Association of Fire Chiefs. He is also the Chairman of the Hazardous Materials Professional Competency Standards of the National Fire Protection Association. He holds a Bachelor's degree in Management from Lewis University. He is a certified Master Instructor through the Office of the Illinois State Fire Marshal and an instructor teaching Fire Science programs for the Chicago City Wide Colleges and the University of Illinois. He has been a member of the Chicago Fire Department since 1969.

P. LAMONT EWELL

P. Lamont Ewell has been the City Manager of the City of Durham, North Carolina since mid 1997. He has also served as Assistant City Manager, Interim Deputy City Manager, and Fire Chief for the City of Oakland, California. Prior to these positions, he served as Deputy Fire Chief, the Interim Director for the County Homeless Shelter, and manager of the Bureau of Emergency Medical Services, Prince George's Co., Maryland.

Mr. Ewell holds a Bachelor of Science Degree in Business Administration from the University of Phoenix and an Executive Masters Degree in General Administration from the University of Maryland. Mr. Ewell has served on the Board of Directors for the National Fire Protection Association, as President of the International Association of Fire Chiefs (1995-1996), and on the Campaign Cabinet on the Board of Directors for the United Way of Greater Durham.

LOUIS T. FIORE

Louis T. Fiore serves as Principal to L.T. Fiore, Inc. in Sparta, New Jersey as a consultant to various companies in the electronic security industry, specializing in product development, product evaluation and communication systems. He has previously served as the Vice-President of Engineering and Purchasing, Vice-President and General Manager of Repco, Inc. of Orlando, Florida, President and Co-founder of Cardinal Technologies, Inc. of New Rochelle, New York, Manager of Product Development at ADT Security Systems, Inc., and at CBS Laboratories, Inc. in Stamford, Connecticut.

Mr. Fiore holds a Bachelor of Electrical Engineering from Manhattan College, a Master of Science in Electrical Engineering from New York University, and a Master of Business Administration from Iona College. He is affiliated with the American Society for Industrial Security (ASIS), the Institute of Electrical and Electronics Engineers (IEEE), and the National Fire Protection Association (NFPA). He is President of the Central Alarm Association (CSAA) and Chairman of the CSAA's Alarm Industry Communications Committee (AICC), a Board Member of the Security Industry Association (SIA) and the National Burglar and Fire Alarm Association (NBFAA).

ARTHUR J. GLATFELTER

Arthur J. Glatfelter has served as the President of the International Association of Fire Chiefs Foundation for the past 20 years and was elected to the Board in 1976. He was previously appointed to the Council of Small Business in the U.S. Chamber of Commerce for four years. Mr. Glatfelter has also served as Board Member, Chairman, and President of the Highway Committee of the York Area Chamber of Commerce between 1972 and 1977. Prior to these positions, he served as President of the York Association of Life Underwriters, the Independent Agents Association of York County, President and current Chairman of the Board of Trustees of the YMCA.

CHIEF RICHARD MARINUCCI

A past president of the International Association of Fire Chiefs (1997-1998), Chief Marinucci has directed the Farmington Hills Fire Department since 1984. The Department has extensive hazardous materials, fire prevention, public fire safety and educational responsibilities, in addition to the traditional fire suppression. He holds three Bachelor of Science degrees, most recently in 1985 from the University of Cincinnati in Fire Safety Engineering Technology.

Chief Marinucci's work at the USFA as temporary Chief Operating Officer until September 1999 concentrated on a refocusing of the Administration's resources toward its core business operations. This included the development of a plan to implement recommendations of the white paper issued by a Blue Ribbon Panel on the USFA, and in advocating a reexamination of the USFA role overall, and the role of the fire service community in particular.

CHIEF JACK K. MCELFISH

Chief Jack K. McElfish has over thirty-five years experience in the career fire service and has served as Fire Chief/Director of Fire and Emergency Services for the City of Richmond since 1995. He has also served as Fire Chief in Clayton County, Georgia (1990-1995) and in Wallingford, Connecticut (1981-1990). In 1998-1999, he served as President of the Southeastern Division of the International Association of Fire Chiefs. In July 1999, he received the International Society of Fire Service Instructors Association "Instructor of the Year" award. He served as an Adjunct Instructor at the National Fire Service Staff and Command Course since 1989 and at the National Fire

Academy from 1980-1990. He has written numerous fire service articles for national fire service magazines and has served as a speaker at over 150 fire/rescue conferences and training programs. In addition, the Total Quint Transformation project he initiated in Richmond received the International City/County Management Association "Public Safety Program Excellence Award" in 1999.

Chief McElfish holds a Bachelor of Science degree in Fire Service Administration with a Minor in Public Administration, a Masters of Science Degree in Fire Science Administration, and a Masters of Arts Degree in Public Administration from the University of New Haven. He also received an A.A. Degree in Fire Science Administration from Montgomery College, Maryland.

DEAN OCKERBLOOM

Mr. Dean Ockerbloom has thirty-nine years of experience in the property-casualty insurance business with the Travelers Property Casualty Company and its predecessors and serves as its Second Vice-President. He has extensive experience in virtually all property and casualty lines of insurance. Mr. Ockerbloom has been responsible for developing commercial property and commercial peril policies both for Travelers and for the insurance industry. He has also created underwriting and loss control standards, training programs and marketing materials for property products.

Mr. Ockerbloom has an Associate Degree in Underwriting and in Risk Management from the Insurance Institute of America. He is also a Chartered Property and Casual Underwriter. Mr. Ockerbloom received a Bachelor of Science in Business Administration from Northwestern University. He has taught at the University of Connecticut School of Insurance. Mr. Ockerbloom serves on the Board of Directors for the Texas Windstorm Insurance Association and the National Disaster Coalition, and a member of the American Insurance Association Property Committee, the Insurance Services Office Commercial Property Committee, and the Insurance Institute of America General Examination and Advisory Committee.

CHIEF BILL PESSEMIER

Chief Bill Pessemier is presently Chief of the Littleton, Colorado Fire Department. He also was the Chief of Rescue Services in Urbana, Illinois and Assistant Fire Chief in Bellevue, Washington. He holds a Masters in Public Administration from the University of Illinois, and has graduated from numerous fire services related educational courses and programs in Washington, Oregon, and the FEMA USFA facility in Emmitsburg, Maryland.

Chief Pessemier's career experience has included the reorganization of regional EMS delivery systems in the Littleton area, using a public/private partnership approach among hospitals, ambulance companies and advanced life support units. He is a member of the National ASTM Committee that votes on national consensus standards for EMS providers. He initiated a community outreach effort called a *Home Fire and Life Safety Survey.* This door-to-door program annually provides residential prevention and life safety information, installation of smoke alarms, or of alarm batteries, free of charge. He also led the reinvention of the University of Illinois' fire service, allowing for elimination of bureaucracy and the improvement of the delivery of fire services.

CHIEF BILL PETERSON

Chief Bill Peterson is the Fire Chief of the Plano Fire Department, Texas. He has also served 25 years as adjunct faculty and active advisor for the fire science program at Moraine Valley Community College in Palos Hills, IL; the College of Lake County, IL; and Collin County Community College, TX. Chief Peterson has chaired the Fire Inspector Professional Qualifications Technical Committee, vice-chaired the Fire Service Training Technical Committee, and served six years on the Standards Council of the NFPA. He also chairs the International Association of Fire Chiefs' Health and Safety Committee, and holds membership at the Professional Development and

Program Planning Committees of the IAFC. Chief Peterson is President of the USA Branch of the Institution of Fire Engineers, has been elected to the International Council of the Institution of Fire Engineers, and is a member of the Society of Fire Protection Engineers. He serves on the Board of Directors of Plano Children's Medical Clinic.

Chief Peterson received his Associate of Applied Science Degree in Fire Science and a Bachelor Degree in Fire Administration from Lewis University. He earned a dual Master's Degree in Public Administration and Human Relations from Webster University. Additionally, he has completed training at the National Fire Academy and in 1985 was a recipient of the FEMA Fellowship to the John F. Kennedy School of Government at Harvard University. In 1997, he received the Benjamin Franklin Leadership Award from the IAFC and Motorola.

CHIEF CAROL SHELTON

Chief Carol Shelton is Chief of the Naval Surface Warfare Center Dahlgren Laboratory (NSWCDL) Fire Department. Since her appointment (1/97), the Fire Department has achieved station and service improvements. Chief Shelton also administers the NSWCDL Emergency Preparedness Program. She recently planned the Department of Defense's Fire and Emergency Services Training Conference (Louisville, Ky, 9/98). Ms. Shelton is a member of the International Association of Fire Chiefs, State Fire Chiefs Association of Virginia, United States Navy and Marine Corps Fire Protection Association.

A second-generation firefighter, Ms. Shelton began as a volunteer firefighter in Fairfax County with Lorton Volunteer Fire Department. She has been in Federal Service since 1980, holding firefighting and safety management positions at NSWCDL, Fort Belvoir in Virginia, Washington, DC, and Quantico Marine Corps Base in Virginia. Ms. Shelton holds an Associate in Applied Science Degree with a major in Fire Science Administration from Northern Virginia Community College. Currently, she is a student at Strayer University in Fredricksburg, Virginia working toward a Bachelor of Science degree in International Business.

CHIEF RONALD JON SIARNICKI

Chief Ronald Jon Siarnicki currently directs and oversees the Fire/Emergency Medical Services Department for Prince George's County, Maryland. The County has the largest combination (volunteer and paid) fire department in the country. Prince George's County has a large and rapidly growing population and, being adjacent to Washington, D.C., often is called upon to participate in some of the special events, and the attendant emerging issues, of present day fire departments. As an example, Chief Siarnicki serves as the Public Safety Liaison between the County government, and the Washington Redskin's National Football League Organization. The emergency operations plan for the team's stadium in Landover Maryland is the Chief's creation.

In addition to the challenges of running the Fire/EMS Department, Chief Siarnicki has also undertaken duties as Assistant State Fire Marshall for the State, and also serves as an adjunct consultant and evaluator fo the U.S. Public Health Service in the delivery of the worldwide National Disaster Medical System. In 1998, he was appointed by the Governor to the Maryland Fire Rescue Education Training Commission, and is consultant for the National Fire Fighter Burn Study, Maryland Fire and Rescue Institute and The Washington Hospital Burn Center. A graduate of the University of Maryland, Chief Siarnicki has over twenty-one years of operational experience. He has received numerous awards from the County, and his most recent professional achievement was the NCCJ Brotherhood and Sisterhood Community Service Award in 1999.

JANE SMITH

Ms. Jane Smith currently serves as the Section Chief of the EMS Academy for the San Francisco Fire Department (SFFD), a position she has held since 1997. She was previously the Paramedic Course Director and a Rapid Zap Instructor for the SFFD. For the Department of Public Health, Safety Division, she served as the Paramedic Captain at Central Medical Emergency Dispatch, Field Training Officer. Ms. Smith was a Primary Instructor (tenured) at the City College of San Francisco.

Ms. Smith has received the Outstanding Contribution to EMS Award (1991) and Paramedic of the Year Award (1997). She is affiliated with several EMS committees including: the Emergency Medical Advisory Committee, EMSA Search Committee (Medical Director), the SFPA (past president), California Paramedic Program Directors (treasurer), Fire Chiefs Association, NHTSA paramedic roll-out (Advisor), and the National Registry Practice Analysis (Advisor).

MELVIN L. STARK

Mr. Melvin L. Stark served as the Senior Vice-President of National and State Legislative and Regulatory Matters and the Manager of Federal Affairs and State Governmental Affairs for the American Insurance Association. Previous to that position, he had extensive experience in claims adjustment, first as a field adjuster and eventually advancing to regional claims manager in both Chicago and Philadelphia. After retirement in 1978, Mr. Stark served as the Federal Affairs Consultant to the Hartford Insurance Group, the American Insurance Association, and other related groups. He managed the Association of Casualty and Surety Companies' Chicago Midwestern Office and worked for the U.S. Fidelity and Guaranty Company. Mr. Stark is a graduate of Baltimore City College and holds an LLB from the University of Baltimore Law School.

ALFRED K. WHITEHEAD

Mr. Alfred K. Whitehead is the General President of the International Association of Fire Fighters (IAFF), an AFL-CIO union representing more than 230,000 firefighters and emergency medical personnel throughout the United States, Canada and their respective territories. Immediately prior to his election as General President of IAFF, he held the position of General Secretary Treasurer from 1982-1988. Mr. Whitehead serves as a Vice-President on the AFL-CIO's Executive Council. He is also a Vice-President of the Transportation Trades Department of the AFL-CIO. Mr. Whitehead was formerly the President of the Los Angeles County Fire Fighters, Local 1014 for more than 12 years and a member for more than 20 years. Mr. Whitehead served as Vice-President of the California Federation of Labor and as Vice-President of the Los Angeles County Federation of Labor before coming to Washington, DC. He began his career as a fire fighter in the Los Angeles county Fire Department in 1954, and departed in 1982 with the rank of Captain.

Mr. Whitehead attended East Los Angeles College and the University of California at Los Angeles. He served in the U.S. Merchant Marine and in the U.S. Army.

CYNTHIA A. WILK

Ms. Cynthia A. Wilk is Assistant Director for the New Jersey Division of Codes and Standards. In her position, Ms. Wilk is responsible for long range planning, the development and implementation of major programmatic initiatives as well as day to day management of Division staff and support activities.

The Division is New Jersey's state construction code agency. The Division implements and oversees the enforcement of a Statewide Uniform Construction Code in New Jersey including building, plumbing, fire protection, electrical, mechanical, barrier free and rehabilitation subcodes. It is responsible for statewide code adoption, supervision of all state and municipal level code enforcement, and training and licensing of local and state inspectors involved in the enforcement of construction codes. Other code related responsibilities include enforcement of the state multifamily housing code, licensing of all rooming and boarding houses, and administration of New Jersey's

New Home Warranty program and the review of all condominium and cooperative sales offerings and administration of many of New Jersey's landlord tenant laws.

Ms. Wilk worked extensively with the Fire Safety Programs with the initiation of the Bureau of Fire Safety in 1985. Working with the State's Training and Education Advisory Board, Ms. Wilk was instrumental on the development and implementation of New Jersey's fire service training and certification program.

She also has been involved with emergency response development in the State. She has worked extensively with the Emergency Management Office of the State Police to develop and define the role of the fire service in responding to emergencies. She coordinated providing hazardous material and Right to Know training to the fire service.

Ms. Wilk serves on the New Jersey Public Employees Occupational Safety and Health Advisory Board and the State Emergency Response Commission. She is the State delegate to the National Conference of States on Building Codes and Standards and serves as the Chair of that Organization's Regulatory Affairs Committee. She has been a member of the Board of Visitors of the National Fire Academy since 1991 serving as Vice Chair 1994-1996 and Chair 1996-1998.

Ms. Wilk is an attorney admitted to practice in the State of New Jersey. She earned her Bachelor of Arts degree at Bucknell University and holds a law degree from Rutgers University. She has been employed by State government since 1974.

Commission Meeting Agendas

1ST MEETING OF RECOMMISSIONED PANEL FOR AMERICA BURNING
SEPTEMBER 15–16, 1999

WEDNESDAY, SEPTEMBER 15, 1999

9:00 AM	*Welcome* - Brian Cowan
	Chairman's Remarks – George Bernstein
	Introductions – ALL
9:30-10:00	*Background on establishment of America Burning ReCommissioned* Rich Marinucci
10:00-11:00	*Panel Members' perspectives* ALL *Identify objective of the Report*
11:00-11:15	*Break*
11:15-11:45	*History of the United States Fire Administration* – Wayne Powell
11:45-12:00	*Questions and Answers about the USFA* – ALL
12:00-12:30 PM	*America Burning: past reports, panels and activities* – Wayne Powell
12:30-12:45	*Questions and Answers about America Burning* – ALL
12:45-1:30	*Lunch*
1:30-3:00	*Initial identification of issues: Individual panel member input* – ALL
3:00-3:15	*Break*
3:15-4:15	*Initial identification of issues, continued* – ALL
4:15-5:00	*Discussion: Panel's approach to issue resolution Ground rules for discussion* ALL

THURSDAY, SEPTEMBER 16, 1999

9:00AM	*Opening* – **Director James L. Witt,** FEMA
9:30-10:00	*Summary of previous day's discussion*
10:00-11:00	*Discussion* – ALL, **Rich Marinucci** (moderator)
	What is the role of the USFA?
	How has the role changed?
	How should it evolve?
11:00-11:45	Discussion – **Brian Cowan** (Moderator)
	The Report of America Burning Re-Commissioned
	Preliminary format/content of final report
	What happens to the report, how is it distributed and to whom?
	What is relationship of USFA and FEMA to final report?
11:45-1:00	Move to Room 273
	Lunch
1:00-1:45	*Congressional Activities Appropriations Authorization* – **Mike Malone and Margaret Larson**
	Questions and Answers – ALL
1:45-3:00	*Discussion of issues as identified during first day* – ALL
3:00-3:15	*Break*
3:15-4:30	*Discussion of issues identified, continued* – ALL
4:30-5:00	*Discussion: Future meetings* – ALL
	Who should be invited to present information to the panel?
	How to manage requests to provide input from non-panel members?
	Appropriate forum for presentations
	ADJOURN

2ND MEETING OF RECOMMISSIONED PANEL FOR AMERICA BURNING NOVEMBER 17-18, 1999

WEDNESDAY, NOVEMBER 17, 1999

8:00 AM	*Continental Breakfast*
8:30	*Welcome and Opening Remarks* – **George Bernstein and Ken Burris**
8:45	*Reports on activities since last meeting*
	Administrative – **Brian Cowan**
	Fire Community input – **John Buckman**
	Data Group – **Ron Coleman**
	"Proactive" Group – **Ron Coleman**
	"Reactive" Group – **Glenn Corbett**
10:30	*Break*

10:45	*Discussion of Fire Data collection and use*
	NDIC – **Steve Worley**
	USFA – **Stan Stewart**
12:30 PM	*Lunch*
1:30	*Panel Discussion: Data Collection and Use* – **Ron Coleman** & ALL
2:45	*Break*
3:00	*Input & Perspectives*
	Volunteer Fire Department – **Fred Allinson**
	Fire Chiefs – **Gary Briese**
4:30 PM	*Open Discussion* – ALL
5:00 PM	*Adjourn*

THURSDAY, NOVEMBER 18, 1999

8:00 AM	*Continental Breakfast*
8:30 AM	*Federal Safety Campaigns and Programs*
	National Youth Anti-Drug Campaign – **Harry Frazier**
	The Office of National Drug Control Policy – **Beverly Schwartz**
9:00 AM	*Buckle Up America*
	Nancy Rubenson, National Highway Traffic Safety Administration (NHTSA)
	Department of Transportation
9:30 AM	*Community Oriented Policing Services (COPS)*
	Timothy J. Quinn, Acting Chief of Staff, Department of Justice.
10:00 AM	*Project Impact: Building Disaster Resistant Communities*
	Kim Fuller, Public Affairs Director, Federal Emergency Management Agency
10:30 AM	*Questions for Presenters* – ALL
11:45 AM	*Discussion: Input from the Fire Prevention* – **John Buckman** & ALL
12:15 PM	*Lunch*
12:30 PM	*Discussion: Final Report Preparation* – ALL
1:30 PM	*Open Discussion: Future Direction of Major Group Activities* – ALL
2:00 PM	*Break into two Groups*
3:45 PM	*Break*
4:00 PM	*Final Discussion and Wrap Up* – ALL
5:00 PM	*Adjourn* – Chairman

3RD MEETING OF RECOMMISSIONED PANEL FOR AMERICA BURNING
JANUARY 11-12, 2000

TUESDAY, JANUARY 11, 2000

8:30 AM   **_Executive Session_**

9:15   **_Opening Remarks_** – Chairman

                 Report and Discussion on Building Codes and Standards

                 Arthur Cote, Senior Vice President for Operations and Chief Engineer, National Fire Protection Association
                 Paul Heilstedt, Secretary-Treasurer, International Code Council
                 Morgan Hurley, Technical Director, Society of Fire Protection Engineers
                 Teresa Deen, Deputy State Fire Marshall – State of Vermont
                 Gerry Jones, Chairman, Multi-Hazard Mitigation Council

11:45   **_Executive Session_**

12:30 PM   **_Lunch_**

1:30   **_Report on Fire Services Leadership Summit_**
                 Chief Luther Fincher, President, International Association of Fire Chiefs

2:30   **_Reports on Federal Support for Fire Services in "non-traditional" areas_**

                 Garry Criddle, Captain, U.S. Coast Guard/U.S. Public Health Service, National Highway Transportation Safety Administration [Emergency Medical Services]
                 Tom Antush, Federal Emergency Management Agency [Terrorism]
                 John Gustaffson, Executive Director, Emergency Response Team, Environmental Protection Agency [Hazardous Materials]
                 (TBD), U.S. Forest Service [Urban/Wildfire Interface]

4:00   **_Executive Session_**

4:30   **_Open Discussion_**

5:30   **_Adjourn_**

WEDNESDAY, JANUARY 12, 2000

8:30 AM   **_Advocating America Burning, Recommissioned_** – **Elisabeth Steele,** FEMA

8:45   **_Current Outline of Panels' Report_** –**William Pessemier**

9:15   **_Discussion and Finalization of Report "Table of Contents"_** – ALL

10:15   **_Discussion of Principal Messages (or Recommendations) of the Panel's Report_** – ALL

11:15   **_Discussion of Principal Messages (or Recommendations) of each Chapter of the Panel's Report_** – ALL

12:30   **_Lunch_**

1:30   **_Break into Chapter Groups_** –ALL

3:30	**Break**
3:45	**Reports of Chapter Groups** – All
5:00	**Next Steps and Adjourn** – Chairman

FINAL MEETING OF RECOMMISSIONED PANEL FOR AMERICA BURNING MARCH 8-9, 2000

WEDNESDAY, MARCH 8, 2000

OPEN SESSION

8:30	**Welcome**
8:45	**Elizabeth Redding,** American Burn Association
9:30	**Dennis Gage,** Insurance Services Office
10:30	**Margaret Neily,** Director Engineering, Consumer Product Safety Commission
11:15	**Margaret Larson** and **Ken Burris** *USFA Authorities and Activities – What Can We Do?*
	Ron Siarnicki *Update Report on Federal Agency Activities to Implement the Original America Burning Recommendations*
11:45	**Lunch**
12:45	**Jack Snell,** Director, Building and Fire Research Laboratory, National Institutes of Standards and Technology
	Richard Bukowski, NIST
	Dan Madrzykowski, NIST

CLOSED SESSION

| 1:45 | **Report Discussion: Development of Recommendations** |
| | **Adjourn** |

THURSDAY, MARCH 9, 2000

CLOSED SESSION

8:30	**Report Discussion**
12:00	**Lunch**
12:45	**Report Discussion, continued**
4:30	**Wrap up**

Identifying Future Challenges Faced by America's Fire Service, Garry L. Briese, Executive Director of the IAFC (presentation)

Discretionary and Mandatory spending: 1966, 1986, 1996

FY 2000 Request to Congress – Operating Accounts

An Example of a Federal Government Highway Safety Education Program – "Buckle Up America"

National Volunteer Fire Summit June 6, 1998

Socioeconomic Factors and the Incidence of Fire, United States Fire Administration

Commission on Fire Accreditation International: Fact Sheets on the Benefits of a Fire and Emergency Service Accreditation Program, Accreditation and Volunteer Fire and Emergency Service Agencies, and *What do I Tell My City/County Manager About Fire Service Accreditation?* Also, List of CFAI Publications and flyer announcing CFAI Online.

COPS: A Ground Breaking Partnership With Local Law Enforcement Celebrates Its Fifth Anniversary, U.S. Department of Justice

An Example of a Federal Government Highway Safety Education Program: "Buckle Up America", Department of Transportation, NHTSA (presentation)

National Youth Anti-Drug Media Campaign: Landmark Social Marketing, Fleishman Hillard International Communications (presentation)

Project Impact: Building Disaster Resistant Communities – Guidebook, FEMA

Risk Watch, National Fire Protection Association

S.1899 (106th Congress, 1st session) – designation of FEMA as Federal Fire and Emergency Management Agency and authorization of grants to fire departments

H.R. 1168 (106th Congress, 1st session) – authorization of FEMA grants to fire departments

Blue Ribbon Panel, Recommendations (1998): Appendices

Daily Fire Report, Thursday, January 6

Making it Work: The Fire Service and SARA Title III, Environmental Protection Agency (fact sheet)

Fire Service Leadership Summit, overhead presentation with letters of support

Excellence: Comprehensive Guide to the IAFC Fire Service Award for Excellence, 1989-1995, International Association of Fire Chiefs (IAFC)

Letter regarding fire prevention programs in the Township of Dover, New Jersey, Lightbody, John,

Codes and Standards for a Safer World, National Fire Protection Association (NFPA)

Brochure on the NFPA as a "partner in code development and option."

Emergency Medical Services Program, January 2000; Overview of EMS Systems in the U.S., National Highway Traffic Safety Administration (NHTSA),

Emergency Medical Services: Agenda for the Future, NHTSA

Emergency Medical Services: Agenda for the Future, Implementation Guide, NHTSA

Protect the Public from Unreasonable Risk of Injury and Death Associated with Consumer Products, U.S. Consumer Product Safety Commission (presentation)

America Burning Recommissioned: Fire Research Issues, National Institute of Standards and Technology (NIST) (presentation)

America Burning and Technology. (NIST presentation)

NIST Fire Research Information for the Fire Services, (NIST presentation)

SFPE Engineering Guide to Performance-Based Fire Protection Analysis and Design of Buildings, Section 3, Overview of the Performance-Based Fire Protection Analysis and Design Process

USDA Rural Development, Fire and Rescue Funding, FY 1999

U.S. National Response Team, (brochure)

Equality and Fairness in the Fire Service: A Thematic Review, HM Fire Service Inspectorate (U.K)

National Engineered Lightweight Construction Fire Research Project – Phase 1, Literature Search and Technical Analysis, National Fire Protection Research Foundation

Solutions 2000: Advocating Shared Responsibilities for Improved Fire Protection, Solutions 2000 Symposium held by the North American Coalition for Fire and Life Safety Education

Life Safety 2000: A Model Plan for the state of Oregon to Reduce the Risk to Oregon Citizens from Fire and Related Emergencies, Oregon State Fire Marshall

Implementing America Burning: The Final Attempt, International Society of Fire Service Instructors

Fire Suppression Rating Schedule, Insurance Services Office

Building Code Effectiveness Grading Schedule, Insurance Service Office

The Wildland/Urban Fie Hazard, Insurance Service Office

Evaluating Building Code Effectiveness: Answers to your Questions, Insurance Service Office

Emergency Response to Terrorism, U.S Department of Justice & Federal Emergency Management Agency

The 2000 Fire Chief/IAFC Guide to Federal WMD Response Assets, 2000 Fire Chief Magazine

Firefighter Fatalities, Washburn, LeBlanc and Fahy, *NFPA Journal*

An Update on Burn Care, Elizabeth Redding, American Burn Association (presentation)

Child Firesetting and Juvenile Arson, Prevention and Intervention - Practitioner's Training Workshop, FEMA/USFA

Child Firesetting and Juvenile Arson ... A Call for Community Action, developed by BurnConcerns for FEMA/USFA (video)

Child Firesetting and Juvenile Arson ... Interviewing Kids at Risk, developed by BurnConcerns for FEMA/USFA (video)

www.ingramcontent.com/pod-product-compliance
Lightning Source LLC
Chambersburg PA
CBHW081222170526
45165CB00009B/2907